T0321649

Second Edition

Modern Physics
and Technology for
Undergraduates

Second Edition

Modern Physics and Technology for Undergraduates

Lorcan M Folan
NYU Polytechnic School of Engineering

Vladimir I Tsifrinovich
NYU Polytechnic School of Engineering

Gennady P Berman
*Los Alamos National Laboratory, and
New Mexico Consortium*

 World Scientific

NEW JERSEY · LONDON · SINGAPORE · BEIJING · SHANGHAI · HONG KONG · TAIPEI · CHENNAI · TOKYO

Published by

World Scientific Publishing Co. Pte. Ltd.

5 Toh Tuck Link, Singapore 596224

USA office: 27 Warren Street, Suite 401-402, Hackensack, NJ 07601

UK office: 57 Shelton Street, Covent Garden, London WC2H 9HE

Library of Congress Cataloging-in-Publication Data
Folan, Lorcan M. (Lorcan Michael), 1960- author.
    Modern physics and technology for undergraduates / Lorcan M. Folan and
Vladimir I. Tsifrinovich (NYU Polytechnic School of Engineering, USA), Gennady P. Berman
(Los Alamos National Laboratory, USA). -- 2nd edition.
       pages cm
    Includes bibliographical references and index.
    ISBN 978-9814723350 (hardcover : alk. paper) -- ISBN 978-981-4730-44-0 (pbk : alk. paper)
    1. Physics. 2. Physics--Problems, exercises, etc. 3. Technology. I. TSifrinovich, V. I.
(Vladimir Ilich), author. II. Berman, Gennady P., 1946–   author. III. Title.
    QC21.3.F65 2015
    530--dc23

                                                        2015027987

British Library Cataloguing-in-Publication Data
A catalogue record for this book is available from the British Library.

Printed in Singapore

Preface

The main purpose of this book is to connect the standard undergraduate introductory course in physics with advances in modern science and technology. The first part of the book gives a very brief review of selected "hot" topics in modern physics and technology. The second part of the book provides problems at the introductory level of undergraduate physics that are related to these hot topics of contemporary science. We emphasize that students can solve these problems with little knowledge of modern physics. They do not have to read the first part of the book. However, if a particular problem stimulates a student's curiosity, he or she can find a brief general tutorial on the subject in the first part of the book. We hope that our book will help to generate some excitement about undergraduate physics and will help students to understand how to apply it to understanding advancing technology.

A huge chasm has developed between modern science and undergraduate education. The result of this chasm is that students who are graduating from college are unable to exploit the many opportunities offered by modern science and technology. Consequently, student interest in undergraduate physics is very low.

We propose to try to bridge this chasm between modern science and technology and undergraduate education. Modern science and technology widely uses the methods of classical physics, but these modern applications are not reflected in the problems assigned to students in most physics courses. Solving practical problems is a very effective way to inform students about contemporary science, to show the important relationships between modern and classical physics, and to prepare students to apply their knowledge in the modern technological environment.

We have prepared a set of problems based on some of the latest developments in science and technology that can be analyzed using the classical physics introduced in standard undergraduate programs.

We did this work in the NYU Polytechnic School of Engineering in cooperation with the Los Alamos National Laboratory (LANL) and the New Mexico Consortium (NMC). The NYU Polytechnic School of Engineering is one of the oldest technical schools in the country, widely recognized as an excellent center for undergraduate education. LANL and NMC are widely known as outstanding research centers, and they have the active programs to promote undergraduates science education.

The authors gratefully acknowledge the assistance of Dr. Gary D. Doolen of LANL, who carefully reviewed the manuscript and provided many corrections and suggestions for its improvement. Any remaining flaws are the responsibility of the authors.

L. M. Folan, V. I. Tsifrinovich and G. P. Berman
March 2015

Contents

Chapter 1

Introduction

A huge gap exists between modern science and technology and standard undergraduate physics education. Typical problems used in undergraduate physics courses often have very little in common with contemporary discoveries and advances. At the same time, many scientists would probably agree that the level of introductory undergraduate physics is sufficient to enable students to estimate and sometimes even compute some important features at the very frontier of modern physics.

The objective of our book is to bridge this gap between modern science and undergraduate physics. We believe that bridging this gap will spark interest in both physics education and the current frontiers of science. We believe that this book will help students to understand and appreciate modern scientific ideas in their future education and working lives.

The book is organized as follows: In the next three chapters, we present a short popular review of selected hot topics in modern science and technology. We hope both college students and college professors will find interesting topics in our review. Chapter 2 presents the current picture of the fundamental elementary particles and fundamental interactions. We briefly describe the three generations of fundamental elementary particles, the compound elementary particles, and the gauge fields. Then we describe the transformations of elementary particles and some of the powerful accelerators designed to study the interactions between the particles. Next, we consider atomic nuclei, nuclear reactions, and quantum properties of atoms including such intriguing topics as Schrödinger cat states. For every topic we provide basic information about the topic as well as current information from the frontiers of contemporary research. The selected topics certainly show some bias and are not intended to be a complete popular review. The same approach is used in the next two chapters.

In Chapter 3, we present some applications of modern physics. We briefly describe the principles of electron spin resonance and nuclear magnetic resonance and the ideas of spin refocusing (spin echo). Then we shift to contemporary ideas of single-spin detection including magnetic resonance force microscopy and scanning tunneling microscopy. We describe nanotubes as one of the frontiers in nanotechnology. Then we discuss superconductivity including Josephson effects and SQUIDs. From these "relatively old" phenomena we shift to frontier problems: superposition states in a nanometer-scale "superconducting box," the Schrödinger cat state for a superconducting current, and superconducting magnets. We also mention applications of physics to fighting natural disasters. As an example, we consider tsunami and contemporary ideas for tsunami warning systems. Finally, we explain the intriguing contemporary ideas of quantum computation and quantum teleportation.

In Chapter 4 we consider astrophysical phenomena. We describe fusion reactions in normal stars, like our Sun. Then we consider the evolution of normal stars. We describe the origins and properties of white dwarfs, neutron stars, and black holes. Finally, we briefly consider some current space flight programs, their achievements and expectations.

Chapters 5 and 6 contain short reviews of the main topics from a standard introductory undergraduate course of physics and the problems. The suggested problems include topics at the frontier of scientific research. Some of the problems are directly connected to the topics discussed in the previous three chapters. To solve these problems a reader does not have to go through the earlier chapters. However, if a topic connected to a problem has excited you, we encourage you to read the corresponding section.

Chapter 5 contains a short review and problems based on an introductory undergraduate, calculus-based course on mechanics, including traveling and standing waves and also sound waves, that are often considered in a separate course.

Chapter 6 contains a review and problems in electricity and magnetism, including electromagnetic waves and even optical phenomena, which are often also treated in a separate course on optics.

Chapter 7 contains hints for solving the problems except for a few very simple problems.

The Appendices contain some useful data and formulas which can be used when solving the problems.

The first edition of this book was published in 2003. Since then a lot of exciting physics phenomena and technological advances have been discovered. Some of them are reflected in our text and physics problems.

In Chapter 2, we have added new sections 2.2 and 2.3. Section 2.2 describes the properties of neutrinos including neutrino oscillations. Section 2.3 is devoted to the Higgs boson. In section 2.7 we have included information about the free-electron x-ray laser and "mini-accelerators," and in section 2.9 we included double electron β-decay.

In Chapter 3 we have added new sections 3.5 and 3.6. Section 3.5 describes the physics of photosynthesis. Section 3.6 is devoted to spintronics and digital magnetic memory. In section 3.4 we have added a description of the 2004 Indian Ocean tsunami. In section 4.2 we have included a description of type Ia supernova which are used as "standard candles" for distance measurements to very distant galaxies. In section 4.3 we have updated information about the International Space Station. Also, we have included information about the Mars Science Laboratory, the Planck and the Kepler Missions.

In Chapter 4 we have added new sections 4.4 and 4.5 describing the search for exoplanets and the mysterious dark matter axions.

Chapter 2

Elementary Particles, Nuclei, Atoms

2.1. Fundamental Elementary Particles

The contemporary scientific picture of the Universe is rather amazing. The Universe is filled with fields. There are three "generations" of fields responsible for the fundamental elementary particles. The "first generation" is the most important. It consists of the "electron neutrino" field, the "electron" field, the "down quark" field, and the "up quark" field. The "disturbances" of these fields we accept as fundamental elementary particles: the electron neutrino (ν_e), the electron (e^-), the down quark (d), and the up quark (u). The most important properties of fundamental elementary particles are: their rest mass, electric charge, and spin.

The rest mass is literally the mass of a particle at rest. When the particle is moving its mass increases, but this effect is significant only when the particle's speed approaches the speed of light. The SI unit of mass is the kilogram (kg). However, for fundamental elementary particles it is more convenient to use the non-SI unit of mass — the electronvolt divided by the speed of light squared (eV/c^2). An "electronvolt" is the energy acquired by an electron when it is accelerated through an electric potential difference of one volt. One eV/c^2 is approximately equal to 1.8×10^{-36} kg.

Measuring the electron neutrino rest mass is one of the great challenges of modern science. It is either extremely small or possibly exactly zero. The rest mass of an electron is approximately 0.51 MeV/c^2 (where M stands for "mega" $= 10^6$). Electrons and electron neutrinos together are called "leptons" (light particles). The rest masses of the up and down quarks are much greater than the electron mass. However, their values have not been measured directly because a free quark has never been observed and current theories predict that single quarks cannot be observed.

The SI unit of electric charge is the coulomb (C). It is convenient to express the charges of fundamental elementary particles in units of the "fundamental charge" (e) which is equal to the magnitude of the electron charge, approximately 1.6×10^{-19} C. Unlike mass, all fundamental elementary particles have values of electric charge which are simple multiples of the fundamental charge. In units of the fundamental charge, the charge of an electron neutrino is zero, the charge of an electron is (-1), the charge of the up quark is $(+2/3)$, and the charge of the down quark is $(-1/3)$.

The spin of a fundamental elementary particle can be thought of very roughly as analogous to the angular momentum of a rotating ball. The SI unit for angular momentum is the joule·second (Js). It is convenient to express the spin of a fundamental elementary particle in units of Planck's constant divided by 2π, which is approximately equal to 1.05×10^{-34} Js. (This unit is commonly designated by the symbol $\hbar$.) In units of $\hbar$, all four particles of the first generation have the same spin, $1/2$. Note that all particles of half-integer spin are called "fermions" after E. Fermi. Particles of integer spin are called "bosons," after S. Bose.

The "second generation" of the fundamental fields and their disturbances (the particles) is much more exotic than the first generation. It also consists of two leptons — the "muon neutrino" (ν_μ) with zero electric charge and the "muon" (μ^-) with electric charge (-1), and two quarks — the "strange quark" (s), with electric charge $(-1/3)$, and the "charm quark" (c), with electric charge $(+2/3)$. The main difference between the two generations is the rest mass of the particles, which is much greater for the second generation. The only directly measured rest mass is that for the muon, approximately 106 MeV/c^2.

The even more exotic "third generation" consists again of two leptons — the "tau neutrino" (ν_τ) with zero electric charge and the "tau lepton" (τ^-) with electric charge (-1), and two quarks — the "bottom quark" (b) with electric charge $(-1/3)$ and the "top quark" (t) with electric charge $(+2/3)$. The only directly measured rest mass is that of the tau lepton — approximately 1800 MeV/c^2.

The complete picture for quarks is a bit more complicated. There are three variants of each quark, called "blue" (B), "green" (G) and "red" (R) quarks. For example, there exist uB quarks, uG quarks and uR quarks, dB, dG and dR quarks, and so on. Thus, we have a total of 18 different quarks.

To complete the list of fundamental elementary particles (and the corresponding fields) we must add the "antiparticles." Each fundamental elementary particle has a corresponding antiparticle of the same mass but with some properties that are "opposite," e.g. electric charge. For example, the electron e^- has an electric charge (-1), and the "antielectron," which is also called the "positron," has an electric charge $(+1)$. That is why a positron is designated as e^+. Electrically neutral particles also have antiparticles. The electron neutrino, ν_e, does not have an electric charge, and an "electron antineutrino," $\bar{\nu}_e$, also does not have an electric charge.

Again, the situation with quarks is more complicated. An "antiquark" not only has the opposite electric charge but it also has "opposite color." As an example, an "up antiquark" $\bar{u}$ can be "antiblue" $(\overline{uB})$, "antigreen" $(\overline{uG})$ or "antired" $(\overline{uR})$. In the same way, there exist three "down antiquarks" $(\overline{dB})$, $(\overline{dG})$ and $(\overline{dR})$. The "color" of the quarks and antiquarks has nothing to do with the visual color of macroscopic objects but represents an intrinsic property like an electric charge. The electric charge of all up antiquarks is $(-2/3)$, and the electric charge of all down antiquarks is $(+1/3)$ — opposite to the charges of the corresponding quarks. Note that all fundamental elementary particles have the same spin, $1/2$. Tables 1 and 2 summarize the properties of the fundamental elementary particles.

2.2. The Strangest Particle

The picture presented in the previous section greatly simplifies the situation with respect to the strangest of the fundamental particles: the neutrino. In order to describe the neutrino's unusual properties we will use a geometrical analogy. Imagine that a unit vector (with the tail fixed at the origin) can be generated with two initial (basis) directions: it points in the positive x-, or in the positive y- direction. The generated vector is not a stationary one: it evolves with time, moving between the +x and +y axes, and from time to time returning

First generation

Name	Symbol	Rest mass (if known) in MeV/c²	Electric charge	Symbol for antiparticle
Electron	e^-	0.511	-1	e^+
Electron neutrino	ν_e		0	$\overline{\nu}_e$

Second generation

Muon	μ^-	105.7	-1	μ^+
Muon neutrino	ν_μ		0	$\overline{\nu}_\mu$

Third generation

Tau	τ^-	1777	-1	τ^+
Tau neutrino	ν_τ		0	$\overline{\nu}_\tau$

Table 1. Fundamental elementary particles: Leptons.

First generation

Name	Symbol	Electric charge	Symbol for antiparticle
Up	u	2/3	$\overline{u}$
Down	d	-1/3	$\overline{d}$

Second generation

Charmed	c	2/3	$\overline{c}$
Strange	s	-1/3	$\overline{s}$

Third generation

Top	t	2/3	$\overline{t}$
Bottom	b	-1/3	$\overline{b}$

Table 2. Fundamental elementary particles: Quarks.

to the initial direction.

Now imagine that you wish to detect the generated vector with a measurement device, which contains the two special detectors, x- and y-detectors. The x-detector detects the vector with the probability P_x if the vector lies on the x-axis. The x-detector does not see the vector if it lies on the +y axis. For any other direction the x-detector detects the vector with the probability $P'_x < P_x$. Similarly, the y-detector detects the vector with the probability P_y if the vector lies on the y-axis. It does not see the vector if it lies on the +x axis. For any other direction the y-detector detects the vector with the probability $P'_y < P_x$. At the time of measurement, the closer the vector to the +x-axis, the greater probability P'_x and the smaller probability P'_y.

The strange vector described above is a simplified analog of the neutrino. Neutrinos can be generated in three basis states, which are called flavors: an electron neutrino, a muon neutrino, and a tau neutrino. The basis state of neutrino is not a stationary one: it smoothly changes with time like a direction of the unit vector in our analogy. An ideal neutrino measurement device contains the three neutrino detectors: the electron neutrino, the muon neutrino, and the tau neutrino detector. The electron neutrino detector can capture an electron neutrino with the probability P. It does not see the muon and tau neutrinos. For any other neutrino state the electron neutrino detector detects the neutrino with some probability $P' < P$. The muon and tau neutrino detectors have similar properties.

Due to the time evolution of the neutrino state the probabilities oscillate with time. As an example, if our measurement device is placed near a particle accelerator, which generates muon neutrinos with a few GeV energy, then the neutrinos can be detected by a muon neutrino detector only, with the probability close to P. If the same measurement device is placed at a distance of 1 Mm from the accelerator, then the neutrinos can be detected by a tau neutrino detector with probability close to P. At the same time the probability for the neutrino to be detected by an electron or muon neutrino detectors will be much smaller than P. At some greater distance from the accelerator, the neutrinos will be detected by a muon neutrino detector with probability close to P. Roughly, it looks like the muon neutrinos change their flavor, then return to the initial flavor again periodically. This phenomenon is called "neutrino oscillations."

Neutrino oscillations are not the only strange property of neutrino. According to quantum mechanics these oscillations indicate that all three neutrino flavors do not have a definite mass. There

are three values of mass m_1, m_2 and m_3, which do not correspond
to the three flavors. It means that if a scientist measures, with high
precision, the electron neutrino mass, he or she will obtain one of
these three values m_1, m_2 or m_3 with some probabilities P_1, P_2 and
P_3, correspondingly. The same is true for the two other flavors. The
only reasonable quantity for any neutrino flavor is the mean mass. So
far, scientists have been unable to measure even the largest neutrino
mass. Experiments show that the mean electron neutrino mass does
not exceed 1 eV/c^2.

There are two other questions which, hopefully, may be answered
in the near future. First, scientists do not know whether neutrinos
and antineutrinos are distinguishable or identical particles. Second,
some experiments indicate existence of a fourth flavor of neutrino,
which is even more elusive than the other flavors. Scientists have even
suggested a name for this flavor: the sterile neutrino. The question
about existence of the sterile neutrino remains open.

2.3.   The Higgs Boson

In the current scientific picture of the world there is one special field,
which is associated with neither the matter particles nor with the
interaction between the particles. This is the Higgs field named after
P. Higgs. The Higgs field fills the whole Universe. It interacts with
the fundamental particles of matter (quarks and leptons). It also in-
teracts with the W^+, W^-, and Z^0 bosons, which mediate the weak
interactions between the fundamental particles. Interaction with the
Higgs field generates the most important property of a particle: its
rest mass. The stronger the particle's interaction with the Higgs field
the greater its mass. Without the Higgs field, the fundamental par-
ticles would have zero rest masses. They would travel with the speed
of light like massless photons. The W^+, W^-, and Z^0 bosons would
also be massless. As a result, the range of the weak interaction would
be infinite like the range of the electromagnetic interaction, which is
mediated by massless photons.

The Higgs field has its own disturbances, which are called Higgs
bosons (H^0). The rest mass of a Higgs boson is about 125 GeV/c^2.
The Higgs boson does not have an electric charge or spin. (A particle

of zero spin is a boson, since zero is an integer). The lifetime of a Higgs boson is extremely small: it is about 10^{-22} s. That is why the Higgs boson can be detected only indirectly by observing the particles generated in the process of its decay. Because the mass of the Higgs boson is so large, it can currently only be produced by the most powerful accelerator ever built by scientists: the Large Hadron Collider (LHC). In the LHC, hydrogen nuclei (protons) accelerated to speeds close to the speed of light move toward each other and collide, producing a lot of "debris" including Higgs bosons.

The most reliable "signature" of Higgs boson production is its decay to two photons with a total energy of about 125 GeV:

$$H^0 \rightarrow 2\gamma. \tag{2.1}$$

A second important "signature" is the decay of H^0 into two Z^0 bosons with the same total energy:

$$H^0 \rightarrow 2Z^0. \tag{2.2}$$

These Z^0 bosons quickly decay into the four leptons (an electron, a muon, and their antiparticles), which are detected in the experiments:

$$2Z^0 \rightarrow e^- + e^+ + \mu^- + \mu^+. \tag{2.3}$$

Both "signatures" have been reliably observed in experiments with the LHC. The 2012 announcement of the first detection of the Higgs boson became a major international event.

2.4. Interactions

According to the current scientific picture, the fundamental elementary particles described above do not interact with each other directly. There exist additional fields, called "gauge fields," which are responsible for the interactions between the fundamental elementary particles. The fundamental elementary particles interact with the gauge fields and, via these fields, with each other. There exist eight "gluon fields," an "electromagnetic field," and three "weak fields" which are responsible for the main interactions between fundamental elementary particles. There is also the "gravitational field" which appears to be much weaker and more complicated than all the other fields.

The gluon fields are responsible for the "color interaction" between quarks. Due to this interaction, quarks of the same color, say red and red, repel each other, and the quarks of different colors, say blue and antiblue or red and green, attract each other. This attraction is so strong that all existing quarks are confined into compound particles, and a free quark has never been observed. That is why the quark–gluon interaction is also called the "strong nuclear interaction."

The electromagnetic field is responsible for interactions between all fundamental elementary particles having electric charge, i.e. all the fundamental elementary particles except the neutrinos. Two particles with the same sign of electric charge, i.e. positive and positive or negative and negative, repel each other, while the particles with opposite signs of electric charge attract each other. At small separations the electromagnetic interaction between quarks is much smaller than the strong nuclear interaction. But the range of the strong interaction is estimated to be quite short, about 10^{-15} m, while the range of the "electromagnetic interaction" is believed to be infinite. For electrons, muons and tau leptons, which do not interact with the gluon fields, the electromagnetic interaction is the strongest interaction.

All elementary particles interact with the weak fields which cause the "weak interaction" between particles. This interaction is much weaker than the electromagnetic interaction and has a range of about 10^{-18} m, i.e. 1000 times less than the strong nuclear interaction. That is why it is very difficult to detect a neutrino, it does not interact with either the gluon or the electromagnetic fields. On the other hand, the weak interaction causes many important transformations. The best known of them is the transformation of a down quark (d) into an up quark (u) with the creation of an electron (e^-) and an electron antineutrino ($\bar{\nu}_e$):

$$d \rightarrow u + e + \bar{\nu}_e. \qquad (2.4)$$

All fundamental elementary particles interact with the "gravitational field." It causes an extremely small attractive "gravitational interaction" between the particles. Normally, the gravitational interaction between two fundamental elementary particles is negligible. However, its range, like the range of the electromagnetic interaction, is infinite and so it is very important to the dynamics of large electrically neutral objects like stars, planets and galaxies.

It would be wrong to think that the gauge fields are only passive mediators of interactions. They have their own disturbances which behave like elementary particles. The best known "gauge particle" is the photon (γ), a disturbance in the electromagnetic field. It has zero rest mass, zero electric charge, and its spin is 1. The three "weak gauge particles" W^+, W^- and Z^0, corresponding to the three weak fields are called the "intermediate vector bosons." The upper index indicates their electric charge ($+1$, -1 and 0). These particles have very large rest masses — approximately 80 GeV/c^2 for W^+ and W^-, and 91 GeV/c^2 for Z^0, (where G stands for "giga" $= 10^9$). The spin of all of these particles is 1. The gauge particles of the eight gluon fields are called gluons, and the gauge particle of the gravitational field is called the graviton. Neither gluons or gravitons have been directly observed.

2.5. Compound "Elementary" Particles

Due to the powerful attraction between quarks of different colors, all existing quarks are confined into particles which are called "hadrons." As nobody is able to split hadrons into quarks, the hadrons are also considered to be elementary particles. There are two types of hadrons — "baryons" and "mesons."

A meson is a bound pair consisting of a quark and an antiquark of "opposite" colors. For example, a pion (π^+) consists of an up quark (u) and a down antiquark ($\bar{d}$) which can be red and antired, or green and antigreen, or blue and antiblue:

$$\pi^+ = (uR, \overline{dR}) \quad \text{or} \quad (uG, \overline{dG}) \quad \text{or} \quad (uB, \overline{dB}) \,. \qquad (2.5)$$

All these color combinations are identical because $\bar{R}R$, $G\bar{G}$ and $B\bar{B}$ cancel out forming a "colorless" pion which is observed in experiments. The electric charge of the pion is the algebraic sum of the electric charges of u and $\bar{d}$:

$$2/3 + 1/3 = 1\,.$$

The symbol "+" in π^+ indicates that the electric charge is $+1$ (in units of the fundamental charge). The spin of a pion is zero. The "antipair," $\bar{u}d$, i.e. $(\overline{uR}, dR)$ or $(\overline{uG}, dG)$ or $(\overline{uB}, dB)$ is designated π^-. It has an electric charge (-1), spin zero and is considered the "antiparticle" of π^+. All mesons are colorless, have an integer electric charge $(0, +1, -1)$ and integer spin $(0$ or $1)$.

A baryon is a bound system consisting of three quarks of different colors. The most important of them — the "proton" (p) — consists of two up quarks (u) and one down quark (d), each of a different color. It may be any of the (uB, uG, dR) or (uB, uR, dG) or (uG, uR, dB) combinations. All three combinations are identical because the three different colors, G, R and B, cancel out, like color–anticolor, forming a colorless proton. The electric charge of the proton is,

$$2/3 + 2/3 - 1/3 = 1\,,$$

and it has spin $1/2$.

The second important baryon — the "neutron" (n) — consists of one up and two down quarks of different colors, (uB, dG, dR) or (uG, dB, dR) or (uR, dB, dG). All three combinations are identical, like the three combinations for the proton. The electric charge of a neutron is,

$$2/3 - 1/3 - 1/3 = 0\,,$$

and its spin is $1/2$.

Similarly, all other baryons are colorless combinations of three quarks. They have a spin $1/2$ and an integer electric charge. All "antibaryons" are colorless combinations of three antiquarks. For example, an "antiproton" $(\bar{p})$ consists of two up antiquarks $(\bar{u})$ and one down antiquark $(\bar{d})$. These antiquarks have different anticolors — one of them is antiblue, one is antigreen and one is antired. The spin of an antiproton is $1/2$, and its electric charge is (-1). Table 3 and Fig. 1 show the properties of a few hadrons.

Name	Symbol	Composition	Rest mass in MeV/c²	Electric charge	Spin	Symbol for antiparticle
Proton	p	uud	938.3	1	1/2	$\bar{\text{p}}$
Neutron	n	udd	939.6	0	1/2	$\bar{\text{n}}$
Pion	π^+	$u\bar{d}$	140	1	0	π^-

Table 3. Selected hadrons.

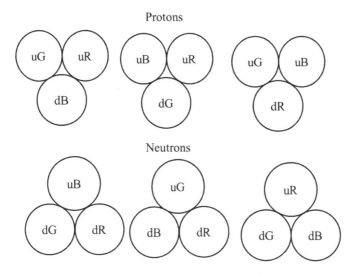

Fig. 1.   Three combinations of up and down quarks form identical colorless protons or neutrons.

2.6.   Transformations of Elementary Particles

An amazing feature of the particle–field interactions is that these interactions cause not only attraction or repulsion between particles, but also the creation (production or emission) and destruction (decay or absorption) of particles. All these transformations are called "reactions" in analogy with chemical reactions. The most common reaction is the spontaneous decay of a particle. There are only two known stable baryons: the proton (p) and the antiproton $(\bar{p})$. All other baryons and all the mesons decay into lighter particles soon after their creation. In all reactions the total energy, momentum, angular momentum and electric charge are conserved. But the number of particles, and even their total rest mass is not conserved!

We mention here a few hadron decay reactions. A free neutron decays into a proton, electron and electron antineutrino:

$$n \rightarrow p + e^- + \bar{\nu}_e. \qquad (2.6)$$

This reaction is the result of the decay of a down quark "hidden" inside the neutron. (See the reaction (2.4).) The mean lifetime of the neutron is about 15 minutes. Similarly, an antineutron decays into an antiproton, a positron and an electron neutrino:

$$\bar{n} \rightarrow \bar{p} + e^+ + \nu_e. \qquad (2.7)$$

Comparing reactions (2.6) and (2.7) one can see that all particles in (2.6) are replaced by their antiparticles and vice versa. This is a general rule: in any formula for a reaction with elementary particles, one can make such a replacement to get a real new reaction.

The pion π^+ has a mean lifetime of only 26 nanoseconds (1 ns = 10^{-9} s), and it decays into an antimuon and a muon neutrino:

$$\pi^+ \rightarrow \mu^+ + \nu_\mu. \qquad (2.8)$$

One can easily write the corresponding reaction for the decay of an antipion π^-.

The most important lepton — the electron — is stable, but the muon and the tau leptons are unstable having mean lifetimes of 2.2 microseconds and 290 femtoseconds, respectively (1 μs = 10^{-6} s, 1 fs = 10^{-15} s). The muon decays into three particles:

$$\mu^- \rightarrow e^- + \bar{\nu}_e + \nu_\mu. \qquad (2.9)$$

The tau lepton has two major decay channels:

$$\tau^- \rightarrow \mu^- + \bar{\nu}_\mu + \nu_\tau, \qquad (2.10)$$

$$\tau^- \rightarrow e^- + \bar{\nu}_e + \nu_\tau.$$

Perhaps the most impressive reaction is the annihilation of a particle and the corresponding antiparticle with the emission of photons. For example, if an electron e^- meets a positron e^+ (its antiparticle), they can annihilate, i.e. completely disappear and generate two photons of high energy:

$$e^- + e^+ \to 2\gamma. \tag{2.11}$$

The same reaction can be observed for a proton p and an antiproton $\bar{p}$:

$$p + \bar{p} \to 2\gamma. \tag{2.12}$$

Annihilation seems to be the only way to convert the total rest mass hidden in elementary particles into electromagnetic energy (photons).

The reaction which is the opposite of annihilation is "pair production," the creation of a "particle–antiparticle" pair. For example, if a neutron n absorbs a photon γ, an up quark (u) up antiquark $(\bar{u})$ pair can be created inside the neutron:

$$\gamma + n = \gamma + (udd) \to udd u\bar{u}. \tag{2.13}$$

This combination decays into a proton p and an antipion π^-:

$$udd u\bar{u} \to (uud) + (\bar{u}d) = p + \pi^-. \tag{2.14}$$

(We do not show all the possible colors — all composite particles are colorless.)

All possible reactions in the world of elementary particles are restricted by conservation laws. Besides the four classical laws — conservation of energy, momentum, angular momentum and electric charge — additional empirical conservation laws have been discovered. The first of these laws, the "Law of Conservation of Baryon Number" states that the difference between the number of baryons and antibaryons remains constant. For example, in the reaction (2.6) we have one baryon (neutron) before the reaction, and one baryon (proton) after the reaction. In the reaction (2.12) we have one baryon p and one antibaryon $\bar{p}$ (the difference is $1 - 1 = 0$) on the left side, and no baryons on the right side. The second empirical law deals with the leptons: the "Law of Conservation of Lepton Number." The difference between the number of leptons and the number of antileptons is conserved. Moreover, this difference is separately conserved for each generation! As an example, in reaction (2.6) there are no leptons before the reaction, and there are a lepton e^- and an antilepton $\bar{\nu}_e$ after the reaction (the difference is $1 - 1 = 0$).

2.7. Accelerators

The best way to study the interactions between fundamental elementary particles is to make them collide with each other. The greater the range of available collision energies for the particles, the more interesting the effects that can be observed. The same is true for the compound elementary particles: studying their collisions is the only way to understand their internal structure. And, finally, to create a new particle whose rest mass is greater than the rest mass of the colliding particles, we have to increase the energy of the colliding particles.

To supply this energy scientists use accelerators — giant factories with highly sophisticated equipment. There are two basic designs of accelerators. The first one is the "linear accelerator" or "linac" where particles move along a straight line, their speed increasing gradually. The linac consists of a set of metal tubes separated by gaps. A particle with electric charge is accelerated by electric fields produced in the gaps. An alternating electric field in the gaps is generated using powerful microwave radiation. The microwave radiation propagates synchronously with the particles, gradually increasing their energy. The longest linac, the Stanford Linear Accelerator (SLAC) in California, has a length of approximately 3 km and can accelerate electrons and positrons to energies of 50 GeV.

By utilizing part of SLAC, scientists created the Linear Coherent Light Source (LCLS) — a free electron x-ray laser. LCLS uses only 1/3 of the SLAC length and accelerates electrons to about 12 GeV. Magnets of alternating polarity change the straight-line motion of the electrons into a zigzag motion, which causes x-ray photons to be produced. The electrons move with a speed close to the speed of the x-ray photons. The photons induce stimulated emission from the propagating electrons without the mirrors required in optical lasers. The powerful coherent x-ray radiation has a wavelength close to the size of an atom. This short wavelength allows researchers to obtain images of tiny objects during the short exposure time of the order of 1 fs (f stands for "femto" $= 10^{-15}$) before the sample explodes under the action of the intense radiation.

The second basic design of accelerators is the "synchrotron." In a synchrotron, charged particles move along a circular path under the action of a magnetic field, produced by powerful electromagnets. When particles pass through regions of applied electric fields, they gain additional energy. The electric and magnetic fields must be strictly synchronized with the motion of the particles to keep them moving on the correct path. The Tevatron, at Fermilab in Illinois, has a circumference of approximately 6 km. It can accelerate protons and antiprotons to energies of 1 TeV (T stands for tera, which means 10^{12}). The Large Electron Positron Collider (LEP) at CERN had a circumference of approximately 27 km. (CERN stands for the research center of the European Council for Nuclear Research, located on the border between France and Switzerland, 100 m beneath the ground.) In the LEP, electrons and positrons collided at an energy of up to 209 GeV. In 2000, LEP was closed down to make way for the Large Hadron Collider (LHC), which is designed to provide collisions between protons with energies up to 14 TeV. There are many other well-known accelerators around the globe. As an example, we mention the Relativistic Heavy Ion Collider (RHIC) at Brookhaven National Laboratory in New York. Its synchrotron has a 3.8 km circumference, and can accelerate protons and other atomic nuclei from deuterium to gold.

Scientists are now developing a linear "mini-accelerator," which could replace giant electron–positron accelerators due to the high rate of the electron energy gain. One way to increase the rate of the energy accumulation is to fabricate two optical gratings. Each grating has a nanometer scale ABABAB... pattern of alternating transparent dielectrics. Electrons propagate in a 400 nm vacuum gap between the gratings (n stands for "nano" $= 10^{-9}$). When an infrared laser beam is focused on the surface of the dielectrics, perpendicular to the direction of propagation of the electrons, the electric field of

the laser beam in the regions "A" is much greater than the field in
the regions "B." The optical gratings are adjusted in such a way
that the electrons travel through the regions "A" when the electric
force points in the direction of motion and the regions "B" when
the electric force is opposite. In this way the infrared beam rapidly
boosts the energy of the electrons.

There are many other important phenomena which can be ob-
served using accelerators. As an example, heavy nuclei colliding at
high energy produce a short-lived "fireball" of quarks and gluons,
called a "quark–gluon plasma." Such collisions are of great scientific
interest.

2.8. Nuclei

While protons and neutrons are "colorless" particles, the gluon fields
provide the strong nuclear interactions between two protons, two
neutrons or between a proton and a neutron. A quark from one pro-
ton (or neutron) inside a nucleus can often interact "slightly" with a
quark from a different nearby proton (or neutron). These interactions
are attractive on average, and at a distance of a few femtometers
(1 fm $= 10^{-15}$ m) this "residual interaction" exceeds the strength
of the electromagnetic interaction (repulsion) between two protons.
The distance between a proton and a neutron in a nucleus is of the
order of 1 fm, and the residual strong interaction at this distance
is estimated to be up to 20 times greater than the electromagnetic
interaction.

Due to this residual strong interaction, protons and neutrons can
attract each other, forming stable atomic nuclei. Nuclei with the
same number of protons but different number of neutrons are called
"isotopes" of the same chemical element. Nuclei with only one pro-
ton are hydrogen (H) nuclei. Hydrogen has two stable isotopes. The
nucleus of the first one is a single proton, p, which is designated as
${}_{1}^{1}H$. The subscript shows the number of protons (atomic number),
and the superscript shows the total number of protons and neutrons
(mass number). The second stable isotope of hydrogen, ${}_{1}^{2}H$, contains
one proton and one neutron. This nucleus has a special name, the

"deuteron." Deuterons can be created when neutrons are captured
by protons:
$$n + {}_1^1H \rightarrow {}_1^2H + \gamma. \tag{2.15}$$

Only 0.015% of all hydrogen nuclei in terrestrial samples are
deuterons. This number, 0.015%, is called the "natural abundance"
(or just abundance) of the ${}_1^2H$ (deuterium) isotope.

The nuclei of all isotopes of the same element have the same elec-
tric charge but, generally, different spin (the intrinsic angular mo-
mentum). For example, the spin of the proton, ${}_1^1H$, is 1/2 (in units
of the Planck constant, $\hbar$). The spin of the deuteron, ${}_1^2H$, is 1. Nuclei
with an even number of protons and an even number of neutrons
have zero spin. All nuclei which have a non zero spin behave like tiny
"permanent magnets." Like all permanent magnets they produce a
magnetic field. The magnitude of this magnetic field is proportional
to the quantity called the "intrinsic magnetic moment" of the nu-
cleus. The unit of magnetic moment in the SI system is Am^2 (ampere
meter squared) or J/T (joule/tesla). 1 Am^2 is the magnetic moment
of a current loop of area 1 m^2 carrying an electric current of 1 A. The
intrinsic magnetic moments of nuclei are normally measured in units
of the "nuclear magneton," $\mu_n \approx 5.05 \times 10^{-27}$ J/T. The magnetic
moment of a proton in these units is $+2.79\ \mu_n$. The magnetic moment
of a neutron is $-1.9\ \mu_n$ and the magnetic moment of a deuteron is
$+0.86\ \mu_n$.

The nuclei which contain two protons are called helium (He) nu-
clei. There are two stable isotopes of helium, ${}_2^3He$ and ${}_2^4He$. The first
one has a spin 1/2. Its natural abundance is only 10^{-4}%. The "main"
isotope, ${}_2^4He$, does not have a spin or a magnetic moment.

The radius of a nucleus (in fm) can be estimated approximately
as 1.25 A$^{1/3}$, where A is the mass number, if the nuclei are assumed
to have an approximately spherical form. The volume of nuclei is
proportional to the mass number, so their density is approximately
constant. It is estimated to be about 10^{17} kg/m^3. For stable nuclei
with mass number up to 40, the number of neutrons is typically close
(or equal) to the number of protons. For heavy nuclei, the number
of neutrons is typically significantly greater than the number of pro-
tons. The heaviest stable nucleus is lead, ${}_{82}^{208}Pb$ (82 protons and 126
neutrons).

2.9.   Nuclear Reactions

The most common nuclear reaction observed is nuclear decay. The majority of nuclei known to scientists are radioactive. Such nuclei are unstable: they quickly or gradually decay into more stable nuclei by emitting particles: electrons (e^-), positrons (e^+), photons (γ), electron neutrinos (ν_e), electron antineutrinos $(\bar{\nu}_e)$, protons (p), neutrons (n), helium nuclei (^4_2He) or even splitting into smaller pieces ("fission"). The lightest radioactive nucleus is the hydrogen isotope ^3_1H which is called "tritium." It decays into ^3_2He by emitting an electron and an electron antineutrino:

$$^3_1H \rightarrow {}^3_2He + e^- + \bar{\nu}_e . \tag{2.16}$$

The emitted electron is called a "β^--particle." The decay itself is called a β^--decay. β^--decay is associated with the weak interaction which causes the transformation of a neutron into a proton inside a nucleus. (See reaction (2.6).) An initial amount of tritium nuclei decreases by half in approximately 12 years. This characteristic time is called the "half-life" of the nucleus.

Many radioactive nuclei exhibit β^--decay. An extremely important example is the β^--decay of the carbon isotope, $^{14}_6C$, into the nitrogen isotope $^{14}_7N$:

$$^{14}_6C \rightarrow {}^{14}_7N + e^- + \bar{\nu}_e . \tag{2.17}$$

$^{14}_6C$ nuclei are constantly being produced in the atmosphere, due to reactions with cosmic rays which strike the atmosphere from space. As a result, approximately 1.3×10^{-12} of all carbon nuclei in the atmosphere are $^{14}_6C$ nuclei. Living organisms constantly take in and release CO_2 to and from the air. Thus, 1.3×10^{-12} of carbon nuclei in living organisms are also $^{14}_6C$ nuclei. When an organism dies, it stops exchanging carbon with the air. New $^{14}_6C$ nuclei do not enter the organism, and the old ones gradually decay away. (The half-life of $^{14}_6C$ is approximately 5730 years.) Thus, by measuring the relative amount of $^{14}_6C$ nuclei, a scientist can find out when the organism died. This is the basis for "carbon dating" of many artifacts made from organic materials.

If a proton inside a nucleus transforms into a neutron, the nucleus emits a positron and an electron neutrino. As an example, the nitrogen isotope $^{13}_{7}N$ decays into the carbon isotope $^{13}_{6}C$, emitting a positron and an electron neutrino:

$$^{13}_{7}N \rightarrow {}^{13}_{6}C + e^+ + \nu_e . \tag{2.18}$$

Such a decay is called a β^+-decay and the emitted positrons are called β^+-particles. This decay is also caused by the weak interaction.

The third type of decay caused by the weak interaction is called electron capture β-decay: a proton inside a nucleus absorbs an electron and transforms into a neutron with the emission of an electron neutrino. The lightest radioactive isotope which experiences electron capture is $^{7}_{4}Be$:

$$^{7}_{4}Be + e^- \rightarrow {}^{7}_{3}Li + \nu_e . \tag{2.19}$$

Electron capture β-decay is possible only if there is an electron somewhere around the nucleus. There thus exists an exciting opportunity to manipulate the rate of the electron capture β-decay by manipulating the electron which is to be captured.

The rarest radioactive decays observed in physics laboratories are the double electron β^--decay and the double electron capture β-decay. The lightest known nucleus, that experiences double electron β^--decay is $^{48}_{20}Ca$. It radiates two electrons and two electron antineutrinos in a single process:

$$^{48}_{20}Ca \rightarrow {}^{48}_{22}Ti + 2e^- + 2\bar{\nu}_e . \tag{2.20}$$

The half-life of this isotope is about 4×10^{19} years. Double electron capture β-decay has been observed with the isotope $^{130}_{56}Ba$:

$$^{130}_{56}Ba + 2e^- \rightarrow {}^{130}_{54}Xe + 2\nu_e . \tag{2.21}$$

Its half-life is about 10^{21} years!

Another important type of nuclear decay is associated with the emission of $^{4}_{2}He$ nuclei, which are called "α-particles." The corresponding decay is called "α-decay." The lightest nucleus which experiences α-decay is $^{5}_{2}He$:

$$^{5}_{2}He \rightarrow {}^{4}_{2}He + n . \tag{2.22}$$

This reaction may also be considered as neutron emission.

All nuclei with atomic number greater than 82 are unstable to α-decay due to the electromagnetic interaction (electric repulsion) between protons. The heaviest long-lived nucleus found in nature is the uranium isotope $^{238}_{92}U$ with a half-life of approximately 4.5×10^9 years. It experiences α-decay into the thorium isotope $^{234}_{90}Th$:

$$^{238}_{92}U \rightarrow {}^{234}_{90}Th + {}^4_2He. \tag{2.23}$$

Nuclear decay is often followed by the emission of a photon or a neutron. The emission of a photon by a nucleus is called "γ-radiation." The photons emitted by nuclei normally have very high energy, often a million times greater than the energy of the photons of visible light.

For a long time, scientists have studied the "transuranic elements": nuclei with atomic number greater than 92. Some of them, e.g. neptunium $^{237}_{93}Np$ and plutonium $^{244}_{94}Pu$, have been found in nature in very small quantities. To produce heavier nuclei, scientists bombard target nuclei with ion beams of lighter or heavier nuclei. In this way they have produced nuclei with atomic number up to 118. Normally, these artificial nuclei are extremely unstable: their half-lives are less than a second. However, there exist relatively stable isotopes even among the transuranic nuclei: one of them, dubnium $^{268}_{105}Db$, has a half-life of approximately 30 hours!

The transuranic elements experience spontaneous fission: the decay of a nucleus into two approximately equal-sized fragments. For uranium $^{238}_{92}U$ the probability of fission is approximately a million times smaller than the probability of α-decay. With increasing atomic number, the probability of fission quickly increases. Fission induced by neutrons is used in nuclear power plants and engines.

Perhaps the most important nuclear reaction is "fusion," which is responsible for the energy production in stars. For example, two hydrogen nuclei ^1_1H (protons) can fuse, producing deuterium ^2_1H:

$$^1_1H + {}^1_1H \rightarrow {}^2_1H + e^+ + \nu_e. \tag{2.24}$$

Deuterium ^2_1H and tritium ^3_1H can combine to produce a helium nucleus ^4_2He :

$$^2_1H + {}^3_1H \rightarrow {}^4_2He + n, \tag{2.25}$$

and so on.

Scientists believe that fusion can be used as a source of energy on Earth. It could become an inexpensive and clean energy source which produces plenty of energy and a small amount of short-lived radioactive waste elements. To use fusion, one must have a fuel at very high temperature so as to overcome the repulsive electromagnetic interaction between the nuclei. If nuclei can approach each other closely enough, the strong nuclear interaction comes into play. At a distance of the order of 10^{-15} m the strong interaction becomes greater than the electromagnetic interaction. The two nuclei combine into a heavier nucleus producing a lot of energy.

The first problem on the long way to controlled fusion is the heating of the fuel to a temperature of the order of $10^8\,°C$. When the temperature of the fuel increases, the electrons are stripped from the atoms. For example, with deuterium–tritium fuel, one will get an overall electrically neutral gas consisting not of atoms or molecules but of independently moving electrons and nuclei, e^-, ^2_1H and ^3_1H. Such a highly ionized gas is called "plasma." Plasmas conduct electric currents, and they can be used to further increase the temperature of a plasma. Electric currents, microwave and laser radiation, even beams of high energy particles — all have been used to further heat plasmas.

The most important problem is how to prevent the plasma from cooling and to maintain temperatures of about $10^8\,°C$. The most natural way is to trap the plasma with a magnetic field, which can keep the electrically charged particles confined inside a small volume of space. There exist many experimental facilities around the globe which are used to implement ideas for a fusion power plant. Probably the most famous of them is the Joint European Torus (JET) in England which can already generate small amounts of fusion power.

2.10.   Atoms — Basic Units of Matter

Due to the electromagnetic interaction, nuclei and electrons attract each other and form bound, neutral atoms. In a neutral atom, the number of electrons is equal to the number of protons in the nucleus.

Thus, the total electric charge of an atom is zero.

The internal properties of an atom differ drastically from the familiar properties of macroscopic particles. The internal energies of an atom take on a discrete set of values. For example, the hydrogen atom, which consists of the hydrogen nucleus (proton) and an electron , can have a minimum value of its internal energy of approximately -13.6 eV. The next possible value of its internal energy is -3.4 eV, and so on. If one measures the internal energy of a hydrogen atom, one will never get the value -10 eV or -5 eV. Normally an atom is found in its ground state — the state of minimum energy. The states corresponding to other energies are called excited states. Together the ground state and the excited states form the set of "stationary states" — the states with definite energies. Using an electric current, or beams of particles, or beams of light, one can transfer atoms from the ground state into excited states. Normally the lifetime of an atomic excited state is very short — much smaller than a second ($\approx 10^{-8}$ s!). Due to the electromagnetic interaction, an atom "jumps" (makes a transition) from an excited state to the ground state (or into an intermediate excited state and then into the ground state) by emitting a photon (or photons). There also exist "metastable" atomic states which do not decay for a relatively long time (from milliseconds to minutes).

The stationary states are only a tiny part of all possible states of an atom. An atom can be in a superposition of two or more stationary states. This means that an atom can be in two or more states of different energies at the same time! It is difficult for us to get a feel for it but we have to accept this fact, because it is based on the results of a large number of experiments. As an example, an atom can be in an equal superposition of two states of energy, E_1 and E_2. If one measures the atom's energy, one will get, with equal probabilities, E_1 or E_2. After the measurement the atom has a definite energy, but before the measurement it did not. Such "superposition states" are of great current interest in physics.

An electron in an atom does not really appear as a point-like particle orbiting a point-like nucleus, but rather as a big negatively charged cloud surrounding a tiny positively charged cloud (nucleus). For different stationary states (i.e. the states with definite energy) the "electron cloud" has different shapes. For a superposition of stationary states, the shape of the electron cloud continuously changes with time.

If an atom has more than one electron, the electron clouds largely overlap. It is then impossible to distinguish the two or more electrons inside the same atom. An atom has a common electron cloud whose electric charge is equal to the sum of the electric charges of all the atomic electrons. Atoms can also combine in various ways to form "molecules." While atoms and molecules are electrically neutral (i.e. the total electric charge is zero), there still are residual electric interactions between them. As a result, atoms and molecules can condense into many familiar liquid and solid substances.

2.11.   Quantum Motion and Schrödinger Cat States

We have already mentioned that an atom does not have to be in a state with a definite energy: it can be in a state which is an arbitrary superposition of stationary states, each with a definite energy. Nevertheless, one normally finds an atom in one of its stationary states: the ground state, the first excited state, the second excited state, and so on. The first question that naturally arises is: why is a superposition, e.g., of the ground state and the first excited state, less likely than either of these two states individually? The answer is connected to interactions with the "environment." The environment includes all the quantum fields in space (whose disturbances we observe as particles) as well as particles themselves. Due to the interactions with the environment, the lifetime of any superpositional state is usually less than the lifetime of a stationary excited state.

The next question that arises is: does this property remain if we consider the very high excited states of an atom? The surprising answer is: no. In this case, the most stable states are not stationary states but rather so-called "quasi-classical states." A quasi-classical state is a superposition of a large number of stationary states which describes a very small, almost point-like electron cloud. Thus, in the limit of very highly excited atomic states, we recover the familiar classical picture of electron motion.

So far we have discussed the interior properties of an atom. What about the motion of an atom as a single particle? If we consider the motion of an atom as a single particle, we typically face the following situation: the atom as a whole unit is usually in a quasi-classical state. It therefore travels like a familiar macroscopic particle.

A most intriguing question now appears: can one produce a super-position of quasi-classical states, e.g. the same atom in two macro-scopically different positions at the same time? It looks like the answer is positive. It is difficult to imagine this, but we have to accept it. Under special circumstances an atom can be transformed into two "quasi-atoms" which travel in two macroscopically different directions. Both quasi-atoms have the same size and structure as the initial atom. Each quasi-atom travels in the same way as if there were no second quasi-atom at all.

And what about a macroscopic particle which consist of thousands of atoms? Can such a particle be transformed into two "quasi-particles" which have the same size and shape as the initial particle? To sharpen this question Schrödinger famously considered a cat which became a superposition of two "quasi-cats," one of them alive and the other one dead. Because of this example, superpositional states involving macroscopic objects or even macroscopically distinguishable trajectories of an atom are called Schrödinger cat states. Incredibly, physics provides a positive answer to the question about macroscopic particles! Recently, Schrödinger cat states were shown to exist experimentally.

The next question that arises: why don't we normally observe such fantastic states in the macroscopic world? The answer again is connected to interactions with the environment. Due to these interactions, the Schrödinger cat state normally decays very quickly. The greater the mass of a particle, the smaller the lifetime of its Schrödinger cat states. The most mysterious and so far unanswered question is the following: how do the Schrödinger cat states decay? If we have two quasi-particles that originated from the same particle, they quickly "entangle" with and become part of their environment. This process is called "decoherence." Then, the environment somehow destroys one of the quasi-particles and transforms the other one into the real particle. The most surprising fact is that this quantum transformation occurs instantaneously even for two quasi-particles which are separated by a significant distance! This very curious property disturbed many physicists. The next question is: what happens if we try to "catch" one of the quasi-particles? The answer is also known. If we try to do this, we artificially strengthen the interaction with the environment. The Schrödinger cat state immediately decays, and we observe a real particle in one of the two possible locations, and nothing remains in the other one.

Chapter 3

Applications of Physics

3.1. Magnetic Resonance

In addition to its electric properties, an electron also behaves like a tiny magnet. To describe the magnetic properties of particles, scientists use the magnetic moment. The same quantity is also used to describe the magnetic properties of a magnetic needle or an electric current loop. The magnetic moment of an electron is very small: it is approximately equal to 9.3×10^{-24} Am^2. This quantity is called the "Bohr magneton," μ_B, after N. Bohr. Note that we describe here the interior intrinsic magnetic moment of an electron which is not related to any orbital motion of the electron. Because of the electron's charge, any orbital motion "generates" an additional orbital magnetic moment, just like a current in a macroscopic current loop.

The intrinsic magnetic moment of the electron is closely related to its intrinsic angular momentum, i.e. its spin. The magnetic moment is a vector quantity, and it is always antiparallel to the electron's spin. The ratio of the magnetic moment to the spin is often called the gyromagnetic ratio. For the electron, the gyromagnetic ratio is approximately equal to the ratio of its electric charge to its mass.

The electronic magnetic moment of an atom can be considered to be the vector sum of the magnetic moments of the individual electrons. If the total electronic magnetic moment is zero, the atom is "diamagnetic." Otherwise the atom is "paramagnetic." The same terms are also applied to ions. The smallest non-zero electronic magnetic moment of an atom is equal to the intrinsic magnetic moment of one electron. If such an atom is placed into an external magnetic field, its ground state (the state with minimum energy) splits into two stationary states: one of them corresponds to the magnetic moment oriented in the direction of the magnetic field, the other one corresponds to the opposite orientation of the magnetic moment.

The energy difference between these two states is proportional to the strength of the magnetic field. The unit of the magnetic field is tesla (T). For example, the Earth's magnetic field is approximately 50 μT (1 $\mu T = 10^{-6}$ T). In a magnetic field of 1 T, the magnetic energy difference between the two states is about 2×10^{-23} joules (J). If we divide this quantity by Planck's constant $h = 6.6 \times 10^{-34}$ Js, we get the frequency of the electromagnetic wave whose photon has an energy of 2×10^{-23} J. This frequency is about 30 gigahertz (1 GHz $= 10^9$ Hz).

If a sample containing many of the paramagnetic atoms considered above is placed into a magnetic field of 1 T, microwaves of frequency about 30 GHz will be absorbed by the sample. The atoms that absorb photons increase their magnetic energy, which is eventually converted into thermal energy. This phenomenon is called electron spin resonance (ESR) or electron paramagnetic resonance (EPR).

There exist a lot of modifications of the basic ESR experiment. One of the most impressive modifications is called electron spin echo. Spin echo can be observed when a sample containing paramagnetic atoms or ions is irradiated by short pulses of electromagnetic radiation of the ESR frequency. After the action of such resonant pulses, the sample emits its own electromagnetic pulse. (See Fig. 2.) This pulse is called the electron spin echo. Its frequency is also equal to the ESR frequency. With well-defined pulse conditions, the shape of the spin echo reproduces the shape of the first electromagnetic pulse, justifying its name "echo." Using spin echos, scientists can study interactions between magnetic moments and the transformation of magnetic energy into thermal energy of a sample.

So far we have discussed the magnetic moment of the electron. The majority of atomic nuclei also have a spin and an intrinsic magnetic moment. The magnetic moment of a nucleus is much smaller than the magnetic moment of an electron. Typically it is expressed in terms of the nuclear magneton $\mu_n \approx 5 \times 10^{-27}$ Am^2. As an example, the magnetic moment of a proton is approximately $+2.8$ μ_n. Even for the simplest nucleus — the proton — there is no simple expression for the gyromagnetic ratio. If a diamagnetic atom, ion or molecule containing a nucleus with non-zero magnetic moment is placed in

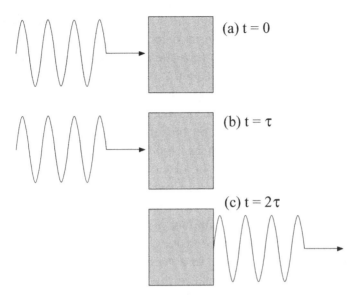

Fig. 2.   Generation of a spin echo. (a) A sample (shown as a rectangle) is irradiated by the first electromagnetic pulse at time $t = 0$. (b) The sample is irradiated by the second pulse at time $t = \tau$. (c) At time $t = 2\tau$ the sample emits an electromagnetic pulse, which is called the "spin echo."

an external magnetic field, its ground state splits into two or more stationary states depending on the magnitude of the nuclear spin. As an example, the ground state of a diamagnetic molecule containing a hydrogen atom splits into two stationary states connected with the two possible orientations of the proton magnetic moment, just like in the case of the electron magnetic moment discussed above. The energy difference between the two states is much smaller than for the electron magnetic moment: in a magnetic field of 1 T it is only 2.8×10^{-26} J. The corresponding frequency of the nuclear magnetic resonance (NMR) is approximately 40 MHz (1 MHz = 10^6 Hz). Just as with electrons, magnetic nuclei can also produce spin echos. While the ESR frequency normally belongs to the microwave region of the electromagnetic spectrum, the NMR frequency typically belongs to the FM radio wave region.

NMR and nuclear spin echo have found many applications in science and technology. The NMR frequency is very sensitive to both the external magnetic field and the structure and environment of a molecule. Thus, NMR can be used for the precise measurement of an external magnetic field and for the study of the molecular structure. Using a sophisticated sequence of electromagnetic pulses and computer analysis of the nuclear spin echo signals, one can get information about the motion of proton magnetic moments in a small region of a human body. The output of this technique is well-known to the general public as magnetic resonance imaging (MRI).

Both ESR and NMR can be observed in a sample containing a macroscopic number of particles. Even in MRI, where the NMR signal comes from a small region of a human body, this small region still contains a macroscopic number of protons. The great challenge of magnetic resonance is to detect it from just a few atoms, and ultimately from a single atom in a sample.

How might one hope to achieve this goal? One of the approaches is called "magnetic resonance force microscopy (MRFM)." To detect ESR or NMR, the MRFM utilizes a very sensitive cantilever — a tiny bar, invisible to the human eye. One end of the cantilever is fixed. The other end can vibrate freely, and has a tiny magnet attached to it. (See Fig. 3.) The cantilever, like a swing, has its own resonant frequency. When the frequency of the spin motion matches the cantilever frequency, the cantilever starts to vibrate sympathetically. Even extremely small vibrations of these cantilevers (of the order of an atomic radius!) can be detected by optical methods, and so MRFM is a very sensitive technique.

Another approach to detecting magnetic resonance utilizes a "scanning tunneling microscope." In this remarkable device, a tungsten probe which has a very sharp tip, sometimes only one atom wide, is placed very close to a conducting surface. Electrons from the conducting surface can jump to the probe providing a small "tunneling" current, typically of a few picoamperes (1 pA $= 10^{-12}$ A). This current is very sensitive to the size of the gap between the probe and the surface: it can detect a change in the gap of the order of 100 picometers, i.e. of the order of an atomic radius. Thus, moving the probe along the conducting surface, a scientist can trace the contours of the surface atoms or ions with magnification up to 10^8.

If a conducting surface contains paramagnetic atoms, and the

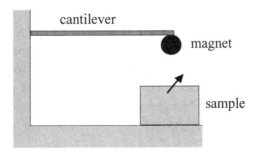

Fig. 3. Magnetic Resonance Force Microscopy (MRFM). A sample (shown as a rectangle) contains a paramagnetic atom. The arrow represents the magnetic moment of the atom. A small magnet is attached to the cantilever tip. The magnet interacts with the magnetic moment of an atom in the sample. This interaction causes the characteristics of the cantilever vibration to change, allowing detection of the atom.

whole sample is placed in an external magnetic field, scientists have observed oscillations (modulations) of the tunneling current. The modulation frequency matches the ESR frequency for the paramagnetic atoms. Because the tunneling current comes from the vicinity of a single atom, such experiments may result in a method for detecting ESR from a single atom.

3.2. Nanotechnology

The building blocks of modern electronics are rapidly shrinking, and research devices are now approaching the atomic scale. Feature sizes of the order of a few nanometers (1 nm $= 10^{-9}$ m) are becoming routine in such research devices. This area of scientific research is often referred to as nanotechnology. It includes many exciting research directions. We will mention only two of them, "nanotubes" and "graphene."

Probably everybody knows two forms of solid carbon: graphite and diamond. Today scientists work with new solid forms of carbon. One of them consists of big molecules which look like tiny cylinders. Such molecules are called nanotubes. This new form of carbon can withstand high stresses and temperatures. The diameter of nanotubes ranges from one to a few nanometers. The length of nanotubes can exceed one micrometer (10^{-6} m). Some nanotubes consist of many coaxial hollow cylinders. Other nanotubes consist of only one cylinder made of one layer of carbon atoms. Nanotubes may become the building blocks of future electronic devices instead of the silicon used in contemporary devices. (The smallest commercial conventional silicon devices have feature sizes no smaller than about 100 nm.)

There are a few different ways to produce nanotubes. For example, scientists may heat a piece of graphite with powerful laser pulses, producing carbon vapor which condenses and forms nanotube molecules. These molecules can then be collected on a surface, such as copper.

What makes nanotubes such a promising material for electronics? First of all, scientists can manipulate the electrical properties of nanotubes by changing their diameters and the number of coaxial cylinders in the tubes. Tubes can be designed to be very good electric conductors, as well as semiconductors. In a semiconductor, electrons cannot move through a sample unless additional energy (e.g. light energy) is transferred to the electrons to excite them into the conducting state. By increasing the diameter of the nanotubes, scientists can smoothly decrease the amount of energy required to put electrons into a conducting state. It is estimated that good conducting nanotubes can withstand an electric current density of up to 10 tera-amperes per meter squared (10^{13} A/m^2), one thousand times greater than the limit for copper.

Another remarkable form of a carbon is graphene — a single atomic layer of graphite. Its surface mass density is less than 1 mg/m^2, which is much smaller than the corresponding value for the thinnest sheet of paper. Graphene can be placed on the surface of various materials, like SiO_2 substrates or metallic films. Surprisingly, scientists have been able to isolate a single layer of graphene, suspending it between metallic bars.

Graphene is the strongest material known to scientists. It is also the best conductor of the electric current at room temperature, with resistivity as small as 10 n$\Omega\cdot$m. (Before discovering graphene the best conducting material at room temperature was silver with resistivity 16 n$\Omega\cdot$m.) Amazingly, graphene is not a conductor, but a semiconductor. The low resistivity of graphene is caused not by a large number of conducting electrons, but rather the high mobility of the electrons. (Electron mobility is the drift speed of the electrons in an electric field of 1 V/m.)

3.3. Superconductors

The phenomenon of superconductivity has been known for a long time, but it is still one of the hottest topics in science. To keep an electric current in an electric circuit, one needs an electromotive force (EMF), e.g. a battery. Without the EMF, the current quickly stops because of the finite electrical resistance of the conductor. An electric current in a conductor is an ordered flow of conducting electrons. The electric resistance originates from the scattering of electrons by thermal vibrations of the positively charged ions or impurities in the conductor. In a typical superconducting material, conducting electrons combine into pairs called "Cooper pairs" after L. Cooper. Unlike an electron which has a spin of 1/2, a Cooper pair has zero spin and is a boson. Bosons have an astonishing property: all of them can be in the same quantum state. Consequently, many Cooper pairs form, producing one big Cooper pair system which spreads out throughout the whole volume of the superconductor. When such a "monster" circulates in a superconducting loop, nothing can stop it, and the associated supercurrent experiences zero resistance.

The Cooper pairs normally appear at very low temperatures. That is why for a long time superconducting properties have been observed only at very low temperatures. Such temperatures are usually quoted in kelvin (K), the number of °C above absolute zero (-273.15°C) where all thermal motions cease. For example, lead becomes a superconducting material at approximately 7 K, mercury at 4 K, and so on. Today scientists and engineers study complicated ceramic materials which become superconducting at temperatures of more than 100 K. As an example, copper oxide containing mercury, barium and calcium becomes superconducting at approximately 130 K.

Very impressive superconducting phenomena can be observed when two superconducting wires are separated by a thin layer of

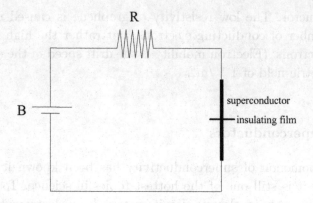

Fig. 4.   A Josephson junction in an electric circuit. B and R are the battery and resistor, respectively.

an insulator. Such a combination is called a Josephson junction, after Brian Josephson. The typical width of the insulating film is of the order of 1 nanometer (10^{-9} m). If a Josephson junction is inserted into an electric circuit, and the electric current in the circuit is less than some critical value, then the Josephson junction does not show any electrical resistance, just like a pure superconductor (see Fig. 4). If the current in the circuit exceeds the critical value, the Josephson junction acquires an electrical resistance. This happens because an alternating current is produced in the junction, and it begins to radiate electromagnetic waves.

The Superconducting Quantum Interference Device ("SQUID") is a superconducting ring containing two Josephson junctions. When such a ring is inserted into an electric circuit (see Fig. 5) it does not show electric resistance until the electric current in the circuit becomes greater than a critical value. Because of "interference" between the two junctions, the value of the critical current shows an oscillating dependence on the magnetic flux through the superconducting ring. Thus, SQUIDs are routinely used for measurement of very weak magnetic fields, in particular, the magnetic fields produced by human tissues.

Because Cooper pairs are all in the same quantum state, scientists study the opportunity to create unusual superpositional states for a macroscopic group of Cooper pairs. As an example, they exper-

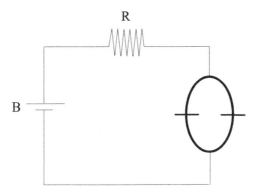

Fig. 5. A superconducting loop with two Josephson junctions inserted into an electric circuit. The notation for the components is the same as in Fig. 4.

iment with Josephson junctions in which one of the superconductors is replaced by a small nanometer-scale "superconducting box." (See Fig. 6.) Using an electric field, scientists can change the number of Cooper pairs in the superconducting box, one at a time. Next, they produce a superposition of two macroscopic states of the box: one of these states corresponds to N Cooper pairs in the box, the other one corresponds to $N + 1$ pairs, where N is a macroscopic integer.

Even more impressive experiments have demonstrated Schrödinger cat states for a supercurrent: the current flows in two opposite directions at the same time. It does not mean that some of the Cooper pairs flow in one direction, and the rest flow in the opposite direction: all the Cooper pairs flow in the two opposite directions at the same time!

The magnetic properties of a superconductor are also extremely interesting. It was believed for a long time that a superconductor can only generate a magnetic moment by electric current flow, i.e. without an electric current the magnetic moment of a superconductor is zero. For almost all known superconductors this statement is true. When a sample of such a typical superconductor is placed in a magnetic field, the magnetic field induces an electric current which flows without resistance along the surface of the sample. This current generates a magnetic moment which points in the direction opposite to the direction of the external applied magnetic field. Thus, if one

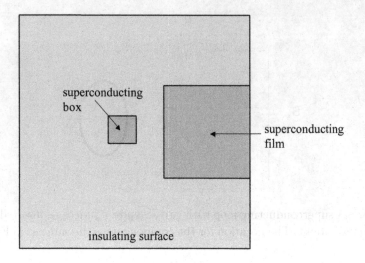

Fig. 6.   A Josephson junction on an insulating surface.

approaches a piece of a typical superconductor with a pole of a magnet, the induced effect is that the magnet and the superconductor always repel each other. This fact is responsible for the well-known phenomenon of "magnetic levitation": a small magnet can levitate above a superconductor cup due to this magnetic repulsion.

Currently, it is known that there exist non-typical magnetic superconductors which have an internal magnetic moment without any electric current, i.e. superconductor magnets! Some of these materials become superconducting at relatively high temperatures, above 50 K.

3.4.   Tsunami

The large domain of applied physics includes not only artificial designs from nanotubes to nuclear thermoelectric generators, but also powerful natural phenomena. A tsunami is one such phenomenon. A wall of water up to 30 m in height and 30 km wide crashing on the shore at a speed of 15 m/s or more may bring death and massive destruction to coastal communities around the globe. Normally, the waves in the oceans are generated by the wind. The length of these waves is relatively short, and the wave motion involves a relatively thin layer of water near the surface of the ocean. The height of the waves can reach up to 30 m but the water far below the surface remains undisturbed.

Typically, tsunamis are generated by underwater earthquakes. They can propagate thousands of kilometers through the oceans with great speeds, up to 200 m/s. With a wavelength of up to 800 km and a height of just a few meters, a tsunami remains almost undetectable in the deep ocean. However this wave involves all the water from the surface to the bottom of the ocean. Near the coast, as the depth of the water decreases, the wavelength and speed of the wave decrease while the amplitude increases significantly.

In spite of the obvious difficulties, scientists are trying to create a reliable tsunami warning system. The system which is known as the Deep-Ocean Assessment and Reporting of Tsunamis (DART) is based on a network of deep-ocean stations. The basic idea is to measure pressure variations near the ocean bottom. While the value of the pressure at the bottom of the ocean is not affected by wind-generated waves, it is sensitive to tsunamis that involve the whole body of water from the surface to the bottom. Information about tsunamis must be converted into sound waves for transmission from the measurement device on the bottom to a buoy on the surface which, in turn, will relay the information to a shore ground station via a satellite using electromagnetic waves. The DART system provides vital information about the direction of tsunami propagation and warns people about the approaching danger.

The 2004 Indian Ocean tsunami became one of the most devastating in human history. It was caused by an earthquake under the Indian Ocean near the Indonesian island of Sumatra. The energy released by the earthquake was estimated to be about 10^{18} J. The duration of the earthquake was about 9 minutes. Ocean water was displaced by the vertical rise of the seabed by several meters. In the open ocean the tsunami wave was less than 1 m high, but it increased to up to 30 m on the coastline near the epicenter. The initial speed of the wave was about 800 km/h, and it moved up to 2 km inland. The energy of the tsunami was estimated to be about 20 PJ (1 PJ = 10^{15} J). The wave reached Sumatra 15 minutes after the earthquake. It affected 14 countries and was detected as far as 8.5 Mm (1 Mm =

10^6 m) away from the epicenter (in South Africa), 16 hours after the earthquake. This Indian Ocean tsunami generated smaller tsunamis in the Pacific Ocean along the western coast of North and South America. In response to the 2004 Indian Ocean tsunami, the Indian Ocean Tsunami Warning System was created.

3.5. Photosynthesis

From the energetic point of view, sunlight is the largest source of energy available on Earth. The rate at which solar energy reaches the Earth (i.e., the power) is approximately 120,000 TW (1 TW = 10^{12} W). This power far exceeds the current energy consumption rate of approximately 20 TW. At the same time, sunlight energy is dilute. The yearly averaged solar power on the Earth's surface is about 170 W per square meter, and it varies depending on geographical location.

All organisms need energy to survive and function. Some organisms, including plants and some protists, bacteria, and blue-green algae, are capable of directly absorbing energy from sunlight. This energy is converted to chemical energy, which is then stored in the form of glucose (sugar). The whole process of energy capture and transfer is called photosynthesis. Photosynthesis takes the energy from solar photons and uses it to build complex molecules. The chemical equation for the photosynthesis can be presented in a simplified form as:

$$6CO_2 + 6H_2O + light \rightarrow C_6H_{12}O_6 + 6O_2.$$

Six molecules of carbon dioxide and six molecules of water, under the influence of sunlight, produce one molecule of sugar and six molecules of oxygen. Finally, the sugar molecules are used to provide energy for the organism. Scientists believe that most of the oxygen in our atmosphere was produced by photosynthetic organisms. Billions of years ago the atmosphere was composed almost entirely of methane, carbon dioxide, and nitrogen.

Chlorophyll molecules absorb sunlight energy and initiate a chain of steps that result in the production of sugar. The initial (primary) processes of photosynthesis occur very quickly (few picoseconds) and include two main steps. The first step is the absorption of mostly red and blue photons by antenna molecules called "light harvesting complexes," LHCs. The photons excite so-called "excitons" in the chlorophyll molecules of LHCs. The excitons move very quickly in the antenna molecules, and reach the "reaction centers" (RCs). In the RCs, electric charge separation takes place. Both LHCs and RCs are embedded in a protein environment. Recently it was discovered that in spite of the room temperature fluctuations of proteins, the primary processes of exciton transfer and electron charge separation are so fast that quantum effects survive and are essential in exciton and charge transfer.

In addition to the electron transfer, protonation is an important part of photosynthesis. Protonation is the addition of a proton (hydrogen ion, H^+) to a system. The concentration, $[H^+]$, of hydrogen ions determines the pH ratio: $pH=-\log[H^+]$, which is an important characteristic of photosynthetic processes and materials. Electrons and protons participate in the creation of electrostatic potentials inside the membrane and in the photosynthetic chemical reactions. The whole photosynthetic engine is very complicated, and differs in different living organisms. So, many fundamental physical, chemical and biological issues need to be resolved in order to understand exactly how this engine works.

This understanding is crucial not only for fundamental science, but for many applications. We mention two of them. One is biofuel production based on unicellular microalgae, which are considered as prospective candidates for producing biofuels. Another application is artificial photosynthesis.

3.6. Spintronics and Magnetic Memory

State-of-the-art computer devices use many types of memories, including dynamic random access memory (DRAM), static random access memory (SRAM) and flash memory. All of these memories are based on electric charge states that dissipate a considerable amount of energy during switching. The main objective of spintronics (spin electronics) is to use the extremely energy-efficient electron spin as information bits to replace the very dissipative charge state bits.

In spite of the tremendous progress of electronic devices, the basic operating principles of these devices remain the same as in electron (vacuum) tubes: controlling an electron current. The principal advance of modern electronics is miniaturization, producing tiny semiconductor transistors that presently switch in a billionth of second and that can be integrated into billion-transistor chips.

In the digital computing industry, electric charge is used to encode digital information (binary bits 0 and 1). However, this approach has a fundamental limitation. Charge is a scalar quantity. Therefore, logic states are defined and quantified by differences in the amount of charge. A fixed amount of charge stored in a device corresponds to the logic bit 0 and a reduced amount of charge (or absence of charge) corresponds to the logic bit 1. This is the approach adopted in the metal-oxide-semiconductor field-effect transistor (MOSFET). When the channel is fully charged, the transistor is "on," encoding the bit, 0. When the channel is depleted of charge, the transistor is "off" encoding the bit, 1.

Switching between logic states requires changing local charge density and the associated energy dissipation, I^2R (I is current and R is resistance). This dissipation is unavoidable in all charge-based electronics. Energy dissipation becomes even more significant when circuit components shrink to nanoscale dimensions, causing heat removal to be a leading problem. Moreover, in nanoscale devices the impurities and quantum effects must be tightly controlled.

Spin, on the other hand, has a fixed magnitude, but a variable direction. In a magnetic field, B, the electron spin has two stationary states — parallel and antiparallel to the field. These two states can encode the logic bits, 0 and 1. In principle, switching between these logic states requires only a spin flip (with no dissipative current). The energy dissipated during a spin flip is approximately the energy difference between the two logic states, 0 and 1, which is the Zeeman energy, approximately $2\mu_B B$, where μ_B is the Bohr magneton. The current state of semiconductor technology is such that this quantum state must be maintained at temperatures less than 1 K.

More recently, the idea of magnetic dots interacting via dipole–

dipole interactions was suggested. In a ferromagnetic dot, approximately 10^4 spins are correlated, and act as a large classical magnetic moment. When the magnetic material has anisotropy, the magnetization direction becomes bistable, and can encode logic bits 0 and 1. This paradigm can work at room temperature. These bits are called Magnetic Quantum Cellular Automata (MQCAs). A very important property of these bits is that the energy dissipated in switching the magnetization of a dot containing $\sim 10^4$ spins is not 10^4 times the switching energy of a single spin, but only about 35 times the switching energy of a single spin.

Spin-Transfer Torque Magnetic Random Access Memory (STT-MRAM) is one of the most promising candidates for next-generation magnetic memory products. The main component of STT-MRAM is the Magnetic Tunnel Junction (MTJ), a sub-100 nm element consisting of two magnetic layers separated by a thin insulating layer (such as MgO). One of these layers is called the reference layer (RL), and the other one is called the free layer (FL). The magnetization of magnetic layers can be either in-plane or perpendicular to the plane. The FL has two energetically equivalent magnetic states, either parallel or antiparallel to the magnetization of the RL. The RL is made so stable that it does not ever switch during any switching operation. Scientists involved in spintronics-based magnetic memory believe that STT-MRAM will be brought to the market in the near future, offering the new functionalities of high-speed, non-volatility and high-density.

3.7. Quantum Computation

Current intense investigations in the field of applied quantum physics have transformed quantum physics itself: instead of the usual stationary states (the states with definite energy) physicists study complex superpositions of stationary states. The "keyword" in contemporary physics has become "information" instead of energy.

The whole idea of quantum computation is based on the use of very sophisticated quantum states. Consider, for example, only two states of an atom. Let the ground state (the state with the minimum energy) be "g," and the excited state with higher energy be "e." According to quantum physics, the atom can be placed in a superposition of these two states. There are infinitely many such superpositions; the simplest of them can be written as (g+e). If you then try to measure the energy of the atom, you will get the energy corresponding to g or e with 50% probability. However, before such a destructive measurement, the atom exists partially in each of the two states, with two different energies, both at the same time — this is the basis for the miracle of quantum superposition!

Now let us consider two identical two-state atoms: atom "1" and atom "2." The simplest superposition state for the two atoms is:

$$g_1 g_2 + g_1 e_2 + e_1 g_2 + e_1 e_2 \,.$$

Here index "1" refers to the first atom, and index "2" refers to the second atom. This state means that we have the two atoms simultaneously in their ground states, in their excited states, and in both combinations of one atom in the ground state and the other atom in the excited state. Everything, and all at the same time! Thus, the two atoms hide $2^2 = 4$ stationary states. Now imagine 1000 atoms in a similar superposition state. They would hide $2^{1000} \approx 10^{301}$ stationary states.

This is often difficult to grasp. A visual aid to help appreciate the complexity of superposition states can be constructed using a thin sheet of clear plastic (e.g. an overhead projector slide). Draw some regularly spaced dark straight lines across the whole length of the slide and then cut it in half and overlay the two pieces. One can observe a whole range of new and different patterns by simply rotating one piece over the other. These Moiré patterns illustrate how two states (patterns) can be superimposed (overlayed) to produce a huge number of new states (patterns). Now imagine that all of these superpositions (patterns) exist simultaneously, and you get some idea of the amazing complexity of superpositional states.

We can assign a definite number to every stationary state. For example, we can use a binary notation like the one used for conventional computers. In binary notation there are only two digits, "0" and "1." The number 2 can be written as (10), the number 3 as (11), the number 4 as (100), the number 5 as (101), and so on. If we assign the digit "0" to the ground state, then $(g_1 g_2)$ corresponds to zero; $(g_1 e_2)$ corresponds to one; $(e_1 g_2)$ corresponds to two, and $(e_1 e_2)$ corresponds to three. Now a superposition of the stationary states means a superposition of these numbers. With only 1000 atoms we can simultaneously represent a huge quantity (10^{301}) of numbers at the same time. Manipulating only 1000 atoms, one could manipulate 10^{301} numbers at the same time. This wonderful opportunity may allow scientists to solve some so-called "intractable" problems which cannot be solved with the most powerful conventional computers, such as the factoring of very large numbers.

Indeed, there is no efficient algorithm which would allow one to decompose the product of two large prime factors. Imagine that one wishes to decompose a 200-digit number using the most powerful supercomputers. It was estimated that the solution of this problem would take about 2 billion years. With a quantum computer the same problem could be solved in a few hours!

There are two very important challenges which have to be overcome on the way to practical quantum computation. The first one is the following: how to extract the hidden information from the atoms. For example, let 1000 atoms hide a huge amount of values of a function. How do we extract the important properties of this function? Currently, a few brilliant quantum algorithms have been suggested which provide solutions to this challenge for some very important intractable problems. The second, more complicated challenge, is the following: how to preserve the fragile superposition states of many atoms and how to correct the inevitable errors. In spite of great progress in the theory of quantum computation, a practical solution to this problem is far from being achieved.

3.8. Quantum Teleportation

Scientists who are working in the exciting area of quantum physics discuss and carry out experiments which remind many of mystical tales or science fiction stories rather than real scientific research. One class of these experiments is on quantum teleportation. The idea of teleportation is the following. Imagine that you have a nice drinking glass and your friend has plenty of small pieces of glass somewhere else. These small pieces could be used to make the drinking glass, but this a very complicated task. We can even assume that nobody knows how to do it. Next, you irradiate the drinking glass with electromagnetic pulses which destroy the drinking glass. You make some simple measurements in the process of destruction. Then, you transfer to your friend the information derived from the measurements. Based on your information your friend irradiates his small pieces of glass with appropriate electromagnetic pulses. As a result, his pieces of glass are transformed into a drinking glass which is an exact copy of your nice drinking glass, while you retain the debris of the original drinking glass. This means that you can interchange the positions of the drinking glass and the small pieces of glass without any knowledge about how to make a drinking glass from the pieces of glass.

This is close to the idea of current experiments on quantum teleportation. The details of the experiments are rather complicated but the main idea is based on utilization of very sophisticated "entangled" states of two particles. The simplest entangled state for two atoms can be written as:

$$g_1 g_2 + e_1 e_2 .$$

This state means that either both atoms, "1" and "2," are in the ground state, or both atoms are in the excited state. There is a rigid correlation (or link) between the atoms: one of them cannot be in the ground state if the other one is in the excited state. It does not look very strange if the atoms are close to each other. However, imagine that these atoms are separated by a large distance! Two atoms separated by a large distance can still be tightly linked to each other. This connection is not the result of some special interaction but is an intrinsic property of quantum systems. If you make a destructive measurement transferring one atom of such a pair into the ground state, you can be certain that at the same moment of time the second atom, which is very far away, will also be immediately transferred into its ground state.

Such a mysterious link between entangled particles can be used to transfer information about the state of a third particle. Suppose, for example, that you wish to transfer the state of a third atom, "x," which is $cg_x + c'e_x$. Here, c and c' are arbitrary complex numbers with only one restriction: $|c|^2 + |c'|^2 = 1$, where $|c|^2$ and $|c'|^2$ are the probabilities for the atom to be found in the ground or excited states in a destructive measurement of energy. You do not need to know the state of your atom to transfer it! Suppose that you already have atom "1," and your friend has atom "2," and these two atoms are entangled. (You should create the entangled state for two initially close atoms and then allow the atoms to move far away from each other, all the time holding the entangled state, which is not a simple task.)

Next, you approach the third atom x with atom 1, so that they can interact to each other. You irradiate the x and 1 atoms with a few electromagnetic pulses and make a destructive measurement of their energy. As a result, the connection between the entangled atoms 1 and 2 will be broken, and atom 2 will be in one of the following four states:

$$cg_2 + c'e_2, \quad \text{or} \quad c'g_2 + ce_2, \quad \text{or} \quad cg_2 - c'e_2, \quad \text{or} \quad c'g_2 + ce_2 \,.$$

These four possible states of atom 2 correspond to the four possible outcomes of your measurement on the atoms x and 1. (You can get $g_x g_1$ or $g_x e_1$, or $e_x g_1$, or $e_x e_1$.) Note: the key information about the numbers, c and c', is already transferred to atom 2! Next, you provide your friend, who is located near atom 2, with information about the results of your measurement. If the result of your measurement corresponds to the state $g_x g_1$, your friend knows that the state of the atom x was transferred to his atom 2. If the result of your measurement is different, say $e_x e_1$, your friend applies electromagnetic pulses which transfer the state $c'g_2 - ce_2$ into the state $cg_2 + c'e_2$. The required pulses are the same for any numbers c and c'. So, your friend can complete the transfer of the quantum state of x without any knowledge about the state itself (i.e. about the numbers c and c'.)

Chapter 4

Space

4.1. Normal Stars

The motions and properties of the planets, moons, stars, galaxies, and even the whole Universe are studied with the application of universal physical laws. Yet, scientists are very far from an understanding of space. Indirect data show that there are many "invisible objects," called "dark matter" which are difficult to study with conventional techniques. Probably the "simplest" objects to investigate are stars, because they radiate visible light. "Normal stars," like our Sun, radiate energy due to fusion of hydrogen ^1_1H into helium ^4_2He in their cores. The sequence of fusion reactions that occur depends on the mass of the star. Scientists believe that for stars whose mass is close to or less than the mass of the Sun, three main reactions are responsible for the energy production. The first reaction is:

$$^1_1H + ^1_1H = ^2_1H + e^+ + \nu_e. \tag{4.1}$$

In this reaction, two protons overcome the repulsive electric interaction between them and fuse (due to their strong nuclear interactions) to form deuterium, ^2_1H, which consists of one proton and one neutron. This reaction occurs in the inner core of a star. As an example, in the core of the Sun the temperature is about 10 million Kelvin (10^7 K). The typical speed for a proton is about a hundred thousand meters per second (10^5 m/s). As a result of collisions, some protons occasionally gain enough energy to approach each other closely enough to actually fuse together. Because the star consists mainly of individual protons (and electrons), it is estimated that about 10^{34} protons fuse every second in the Sun's core. As a result of this reaction, the kinetic energy of the particles increases. The positron, e^+, which is the antiparticle of the electron, quickly "annihilates" with an electron producing two γ-ray photons which, in turn, will be scattered by other particles, increasing their kinetic energy.

The second reaction is:

$$^2_1H + ^1_1H = ^3_2He + \gamma. \tag{4.2}$$

Here a proton ^1_1H and deuteron ^2_1H fuse together producing the helium isotope, ^3_2He, and emitting a γ-ray photon. The third reaction,

$$^3_2He + ^3_2He = ^4_2He + ^1_1H + ^1_1H + \gamma, \tag{4.3}$$

produces the helium nucleus, ^4_2He, and two protons ^1_1H . Due to these three reactions, which are called the "proton–proton chain," the number of helium nuclei, ^4_2He, increases with time.

Scientists believe that the proton–proton chain is responsible for the huge luminosity (light power emitted) of the Sun. At the same time, the fusion reactions support the structural stability of the normal star. Due to gravitational attraction, the star's particles are pulled toward the center of the star. Because of the fusion heating in the core of the star, the pressure increases, forcing the particles outward from the core. These two processes reach an equilibrium where they are in balance with each other, providing stability for a normal star. The luminosity of a normal star quickly increases with increasing mass. Correspondingly, the greater the mass of a normal star the shorter its lifetime. The mass of our Sun is approximately 2×10^{30} kg. Its luminosity is estimated as 4×10^{26} W, and its lifetime is expected to be about 10 billion (10^{10}) years. The current age of the Sun is estimated to be about 4.6 billion years.

4.2. Unusual Stars

Today many "unusual stars" are known, whose properties are very different from the properties of normal stars. Scientists believe that many of these unusual stars originated from normal stars, as a result of their natural evolution.

One type of well-known unusual star is the "White Dwarf." Their mass is about 60% of the Sun's mass, but their radius is only about

1% of the Sun's radius, i.e. close to the radius of the Earth! (The radius of the Sun is about 700 megameters (7×10^8 m), and the radius of the Earth is 6.4 megameters). The surface temperature of a White Dwarf is greater than that of the Sun, but their luminosity is typically 100 times less than the Sun's luminosity. Scientists believe that White Dwarfs originated from normal stars whose mass was close to, or smaller than, the Sun's mass.

The main idea of a White Dwarf's origin is the following. Eventually, most of the protons in the core of a normal star are converted into helium nuclei. The rate of the fusion reaction drops, and it cannot maintain the stability of the star's core anymore. So, the star's core shrinks in size due to the gravitational attraction. As a result, the kinetic energy of the helium nuclei (and, consequently, the temperature of the core) increases. When the temperature in the core reaches about 100 million Kelvin (10^8 K) fusion of helium nuclei becomes possible. As a result of this fusion, helium nuclei are converted into carbon nuclei, increasing the kinetic energy of the particles and emitting γ-radiation. The star again becomes stable and continues to radiate light. At a temperature about 1 billion Kelvin (10^9 K) carbon becomes a fuel, and so on. The fusion of nuclei less massive than iron can produce additional energy to heat the core of the star. The fusion of the iron and more massive nuclei requires energy and consumes some of the star's energy.

The change from hydrogen fuel in the core to helium fuel is accompanied by a change in the star's radius and color. The overall radius increases, and the star's color becomes red. The star becomes a "Red Giant." When the helium fuel in the star's core is exhausted, the star loses its outer layers ("envelope"), and the star's core contracts to a small size.

The next force that can fight the gravitational contraction is associated with the quantum properties of electrons. Electrons oppose the decrease in the star's volume not because of the electrostatic repulsion between them (as the number of electrons and protons in the star is approximately the same the star is electrically neutral) but because of their half-integer spin: two identical particles having half-integer spin cannot occupy the same quantum state. This law is called the "Pauli Exclusion Principle" after Wolfgang Pauli. If the mass of the star is not very big, the pressure associated with the "electron gas" can withstand the gravitational attraction, producing a new but smaller equilibrium size. Such a star is observed as a White Dwarf. The White Dwarf slowly radiates the energy collected in the process of gravitational contraction and eventually becomes dim. Unlike a normal star, the greater the mass of the White Dwarf, the smaller its radius.

If the mass of the star's core is greater than 1.4 times the Sun's mass, the electron gas cannot withstand the gravitational attraction. Such massive stars normally explode, ejecting their envelopes and leaving the core, which contracts to a very small size. Such an explosion is called a "supernova," whose peak luminosity is comparable to the luminosity of a whole galaxy! In 1987 scientists detected a supernova explosion very close to our own Milky Way galaxy, in the Large Magellanic Cloud which is a satellite galaxy of the Milky Way. An extremely faint star, which could be barely observed with a telescope, in one day became visible to the naked eye. This supernova explosion was also recorded by underground neutrino detectors.

If the electron gas in the star's core cannot resist the gravitational attraction, the nuclei of the core break apart and the protons start to absorb electrons. In this reaction, protons p and electrons e^- transform into neutrons n and emit electron neutrinos ν_e:

$$p + e^- \to n + \nu_e. \tag{4.4}$$

This process gives birth to the second well-known type of unusual stars, the "neutron stars." In a neutron star the "neutron matter" resists the gravitational attraction, again because of the Pauli Exclusion Principle (the neutrons, like electrons, have spin $1/2$). The radius of a typical neutron star ranges from 10 to 15 km. The greater the mass of the neutron star, the smaller its radius. So far it is unclear what the maximum possible mass of a neutron star is.

A typical neutron star rapidly rotates about its axis. Its period of rotation ranges from a few milliseconds to a few seconds. Scientists believe that "Pulsars," which radiate periodical electromagnetic pulses detected by radio-astronomers, are neutron stars. The time interval between neighboring electromagnetic pulses is believed to be equal to the period of the neutron star's rotation. From observations of pulsars, scientists were able to measure the masses of neutron stars. They range up to approximately two solar masses.

If the mass of the star's core after the supernova explosion exceeds the maximum mass possible for a neutron star, the "neutron matter" cannot resist the gravitational contraction. The gravitational contraction then gives the birth to the much discussed but very unusual and mysterious object called a "Black Hole." These are believed to be invisible objects, which absorb any radiation or matter approaching them. A Black Hole is the source of a very strong gravitational field. Many stars (maybe even a whole galaxy) can orbit around a Black Hole, like planets orbit a normal star.

Often two stars form a binary system, where they orbit their common center of mass. If one of the stars is a White Dwarf, then the evolution of the binary system may result in a spectacular explosion, which is called type Ia supernova. This phenomenon can be explained by assuming that the White Dwarf contains a lot of carbon, and that matter from the binary companion star is gradually transferred to the White Dwarf. Then the temperature in the core of the White Dwarf increases to the critical point for carbon fusion. Carbon nuclei can then fuse together producing magnesium nuclei in excited states:

$$\ _{6}^{12}C + \ _{6}^{12}C \rightarrow \ _{12}^{24}Mg^{*}. \tag{4.5}$$

(The asterisk on the right side of this equation denotes an excited state of magnesium.) The magnesium nucleus $_{12}^{24}Mg^{*}$ "jumps" to its ground state radiating a photon, or decays to other nuclei, for example:

$$\ _{12}^{24}Mg^{*} \rightarrow \ _{10}^{20}Ne + \ _{2}^{4}He. \tag{4.6}$$

Carbon fusion generates a huge amount of radiation. The whole star explodes like a cosmic thermonuclear bomb. The power generated by a type Ia supernova has a universal characteristic dependence on time. This remarkable property is used to measure the distance to very distant galaxies, which host type Ia supernova.

4.3. Space Flights

Space flights have become a routine part of our lives. But still they have growing importance for physics, technology and space exploration itself. One of the most important research laboratories is the International Space Station (ISS), which orbits the Earth at a height of about 360 km above the ground. Scientists are conducting a lot of experiments in the weightless conditions on board the ISS.

The first laboratory module on the ISS — NASA's research station, Destiny — was activated in 2001. It has a cylindrical form, more than 8 m long and 4 m in diameter. Subsequently, four other laboratory modules have been activated: the Japanese Experiment Module (Kibo), the Columbus module of the European Space Agency (ESA), and two Russian modules, Poisk and Rassvet. Columbus carries the Solar Monitoring Observatory (SOLAR), which studies the Sun's radiation in the wavelength region between 17 nm and 100 μm (ultraviolet, visible, and infrared parts of spectrum). Kibo carries the Monitor of All-sky X-ray Image (MAXI), which detects x-ray radiation in the region of wavelengths from 42 pm to 2.5 nm. Every 96 minutes MAXI provides a full sky survey of x-ray sources.

The Alpha Magnetic Spectrometer (AMS-02) installed on the ISS in 2011 detects electrically charged particles traveling through space. It has a permanent magnet with unlimited lifetime, which produces a magnetic field of about 150 mT, and a cryogenic superconducting magnet with a 3-year lifetime and magnetic field about 750 mT. Data from AMS-02 indicated a surprising phenomenon: the ratio of number of positrons to the number of electrons observed gradually increases in the energy range from 10 GeV to 250 GeV.

The Planck mission (the Planck space observatory) launched in 2009 detected electromagnetic radiation in the range of wavelengths from 350 μm to 10 mm. In particular, it provided a very accurate map of the cosmic microwave background — the oldest radiation in the Universe generated long before the formation of the first stars. The Kepler mission (the Kepler space observatory) was also launched in 2009 to search for planets orbiting other stars in the Milky Way galaxy. Using the data obtained by the Kepler observatory, scientists have found Earth-like planets orbiting Sun-like stars where life may exist.

A fleet of solar spacecraft help scientists to study the "Sun's Envelope." While the nuclear reactions in the Sun's core have been explained, the outer part of the Sun remains a "puzzle" for scientists. The visible surface of the Sun is called the "photosphere." Above the surface region is the "chromosphere," and above that the "corona." The corona radiates the "solar wind" — the outward flow of electrically charged particles from the Sun. Every 11 years the Sun's surface becomes more violent, producing firestorms which cause distortions in satellite and even ground-based communications and power distribution systems. The temperature of the photosphere is about 6000 Kelvin. Surprisingly, the temperature of the chromosphere is higher, about 10,000 Kelvin, and the temperature of the corona is even higher than the temperature of the chromosphere: it reaches about 1,000,000 Kelvin!

Solar spacecraft are expected to help resolve the temperature and other puzzles of the Sun's corona. They detect visible, ultraviolet, and x-ray radiation from the Sun. As an example, the Solar and Heliospheric Observatory (SOHO) satellite was launched into a special "halo" orbit that keeps the satelite on a direct line between the Earth and the Sun, allowing it to continuously monitor the Sun's radiation without obstruction.

A number of other spacecraft study the mysteries of the planets, their moons, asteroids and comets orbiting the Sun. For example, data from the Galileo probe indicated the existence of liquid oceans under the surface of one of Jupiter's moons. Scientists suspect that these oceans could support some form of life.

The Near Earth Asteroid Rendezvous (NEAR) spacecraft landed softly on the asteroid Eros at a distance of about 300 billion meters from the Earth. The photos obtained by NEAR showed that the surface of the asteroid is rocky. NEAR's instruments also detected γ-radiation emitted by the asteroid allowing analysis of the composition of the asteroid.

In 2012 the Mars Science Laboratory (MSL) delivered a 900-kg mobile rover called Curiosity to the surface of Mars. The six-wheeled rover can move through soft sand and over solid rock with speeds up to 90 m/hr. Data received from Curiosity will be important for future manned missions to Mars.

The space-based Chandra X-Ray Observatory, Multi-Mirror Mission and other x-ray satellites have detected the most violent events in the Universe: supernova explosions of stars, absorption of matter by Black Holes, and even collisions of stars. Cosmic x-rays cannot reach the Earth's surface and may only be studied from spacecraft. X-ray astronomy also indicates the existence of "magnetars" — neutron stars with the greatest known magnetic field, up to 100 gigatesla (10^{11} T). Chandra has also detected the signatures of massive Black Holes at the centers of many galaxies.

We have certainly mentioned just a few recent space flight missions. Many other exciting space exploration programs are on the way.

4.4. Exoplanets

We are currently in the midst of a "golden age" of planetary discovery and characterization. While our solar system's planet count has fluctuated historically in the single digits, the count of extrasolar planets or exoplanets (those orbiting stars other than the Sun) has exploded from zero to more than a thousand in just the past few years. The likelihood that other habitable planets exist orbiting other stars has been the stuff of science fiction for centuries, but recent advances have brought us to the brink of identifying and characterizing other planets very much like our own.

The conceptually simplest detection method is direct imaging, i.e., using a telescope to directly observe planetary objects orbiting other stars. While it sounds simple, it is an extremely difficult undertaking. All the target stars are very far away (by terrestrial standards), and since any planet would be relatively small and close to its star, they are difficult to resolve optically. A number of techniques have been developed to reduce the obscuring influence of the star, from placing an obstruction at the star's image in the focal plane of the telescope, to using interference effects to cancel or null the star, to using infrared radiation rather than reflected star light to make images. Several nearby stars including Fomalhaut and HR 8799 have directly visible planets. Future enhanced observations will likely soon expand this list considerably.

While being able to directly "see" the new planets is wonderful, their presence can often be detected even though they are too far away to directly image. One successful detection method has been developed based on astrometry, the very precise measurement of the positions and true or proper motions of the stars. Until the early 19th century, measurements indicated that the positions of the stars were fixed and constant with respect to each other. The sizes and shapes of the constellations appeared unchanging for millennia and the rotation, revolution, precession and other motions of the Earth explained the periodic movements of the stars observed in the night sky. However, once it became possible to detect the very gradual changes in the relative positions of the stars, it was discovered that some stars are gravitationally bound to each other and are members of binary or multiple star systems. The members actually orbit the center of mass of the system and the motions can be analyzed using Newton's Laws. If the masses of the stars are very different, then the largest one may appear to "wobble" periodically rather than execute an orbit. Once the measurements became precise enough, it was possible to detect the wobbling produced by smaller, non-stellar members of a system, i.e., planets. The earliest to be detected were large gas planets, orbiting close to the host star. These have come to be known as "Hot Jupiters" because they resemble "our" Jupiter but have much smaller orbits and higher surface temperatures.

Other developments in solar research allowed the detection of resonant mechanical oscillations of the Sun, caused by various types of internal wave motions. These oscillations are readily detected using the periodic Doppler shifts they produce in the spectral lines of the Sun. This helioseismology technique has identified hundreds of normal vibrational modes and analysis of them has provided detailed information about the interior structure of the Sun. Extension of the technique to other stars soon followed and some very long period oscillations were observed. These turned out to be due to wobbling of the star about the center of mass of its system in response to the motions of nearby planets. Like astrometry, this radial velocity method is most sensitive to larger planets orbiting close to the star, so the early exoplanet catalogs featured mostly Hot Jupiters.

Another method led to a recent large increase in the number of confirmed exoplanets. It is based on the dimming of a star when a planet passes in front of it, on a direct line between the star and the observer. These planetary transits produce a characteristic brightness reduction, and the depth of the minimum is directly related to the relative sizes of the star and the planet. This method requires an unlikely alignment of the planetary orbit and the line of sight to the observer. But if a large number of stars are monitored, the probabilities accumulate and a significant number of examples can be expected. The dedicated robotic spacecraft, Kepler, was launched to make systematic measurements of over a hundred thousand stars. During its lifetime Kepler produced thousands of planetary candidates and a planned follow-up mission called the Transiting Exoplanet Survey Satellite (TESS) is due to be launched in 2017. Kepler focused on a particular patch of sky and showed that the transit method works well. TESS will instead systematically search our galactic "backyard," seeking planetary systems around the stars nearest to the Sun.

A number of other planetary detection methods are being developed and the range of planetary characteristics being observed is remarkable. Our solar system is still unique, but it is now known to be only one of many possible configurations of planets in orbit about a star.

4.5. Axions

During the last few decades, our understanding of the matter in the Universe changed significantly. In the "old" picture of the Universe, protons, neutrons and electrons dominated. In the "new" picture, dark matter makes up about 85% of all the matter in the Universe. For decades, physicists have struggled to unravel the mysteries of dark matter. Recently, three candidates for the dark matter particles have been widely discussed; sterile neutrinos, weakly interacting massive particles (WIMPs), and axions. Some scientists believe that axions are the main constituent of dark matter. They estimate that every cubic centimeter of space could contain about 100 trillion axions, produced during the Big Bang about 14 billion years ago. It is expected that axions have no electric charge, no spin, an extremely small mass, $(10^{-12} - 10^{-9}$ of the electron mass) and very weak interactions with ordinary matter. Consequently, detection of these particles requires extremely sensitive measurement techniques.

The search for axions conducted in the Axion Dark Matter eXperiment (ADMX) relies on the fact that the axion's interaction with a large magnetic field in a resonant cavity may produce microwave photons. The ADMX detector is a meter-scale cylindrical electromagnetic cavity in a static magnetic field, $B = 8T$. When axions entering the cavity are in resonance with an electromagnetic resonant mode of the cavity, they can convert the axions into photons with the frequency of that cavity mode.

These experiments started more than 30 years ago. The main problem is to detect the very weak microwave radiation produced by the axions. To do this, researchers have constantly improved the sensitivity of detectors. An ultra-sensitive microwave receiver, which uses superconducting electronics, amplifies the cavity signal to the point where spectral analysis can be used to search for signatures of the photons expected to be emitted by axions.

During the past 30 years, the sensitivity of detectors have improved by at least a factor of 400. The first detectors operated at a temperature $T \approx 1.5K$. Future experiments are expected to be performed at temperature of $T \approx 100mK$. This upgrade will improve system performance to such an extent that ADMX will be sensitive enough to detect axions (or be able to rule them out) as a component of our galaxy, for all plausible coupling strengths.

Recently, scientists reported analysis of the 12-year data set collected by the European X-ray Multi Mirror Mission (XMM-Newton). They came to conclusion that the x-rays they detected were generated in the magnetic field of the Earth, and are consistent with a steady flow of axions, created in the Sun's core. Future space-based experiments are expected to confirm this conclusion.

Chapter 5

Mechanics

5.1. Review

1. Classical Mechanics starts from the description of the motion of a "particle." A particle is a small, point-like, but macroscopic rigid solid entity. It contains many atoms but its physical size is negligible in problems under consideration.

2. The "position" of such a particle can be described using a position vector $\vec{r}$, which connects the origin of a Cartesian coordinate system with the particle. The vector $\vec{r}$ is then described in terms of Cartesian coordinates of the particle

$$\vec{r} = \vec{i}x + \vec{j}y + \vec{k}z , \qquad (5.1)$$

where $\vec{i}$, $\vec{j}$, and $\vec{k}$ are "unit vectors" in the positive x, y, and z directions. The displacement of a particle $\vec{\Delta r}$ is given by

$$\vec{\Delta r} = \vec{r}_f - \vec{r}_i , \qquad (5.2)$$

where $\vec{r}_i$ and $\vec{r}_f$ are the initial and final positions of the particle, respectively.

3. The "velocity" of a particle $\vec{v}$ is the rate of change of its position

$$\vec{v} = \frac{d\vec{r}}{dt} = \vec{i}v_x + \vec{j}v_y + \vec{k}v_z . \qquad (5.3)$$

4. The "acceleration" of a particle $\vec{a}$ is the rate of change of its velocity

$$\vec{a} = \frac{d\vec{v}}{dt} = \vec{i}a_x + \vec{j}a_y + \vec{k}a_z . \qquad (5.4)$$

5. The "average velocity" is defined as the average rate of change of
the position

$$\vec{v}_{av} = \frac{\vec{r}_f - \vec{r}_i}{\Delta t}, \tag{5.5}$$

where $\vec{r}_i$ and $\vec{r}_f$ are the initial and final positions in a problem under
consideration, and Δt is the time interval of the motion.

6. The distance s traveled by a particle is the length of its trajectory,
i.e. the path actually traveled by the particle.

7. The speed of a particle is the rate of increase of the distance
traveled

$$\text{speed} = \frac{ds}{dt}. \tag{5.6}$$

The speed of the particle is equal to the magnitude of its velocity $\vec{v}$.

8. The average speed of a particle is defined as the ratio $s/\Delta t$. In
general, the average speed is not equal to the magnitude of the av-
erage velocity because of the vector nature of a finite displacement.

9. Any two particles affect each other through "interactions." If Par-
ticle 1 interacts with Particle 2, then Particle 1 exerts a "force" $\vec{F}_{12}$
on Particle 2, and Particle 2 exerts a force $\vec{F}_{21}$ on Particle 1.

10. The "Action–Reaction Principle" ("Newton's Third Law") states
that $\vec{F}_{12} = -\vec{F}_{21}$.

11. The net force "$\vec{F}_{net}$" acting on a particle is equal to the vec-
tor sum of all the forces acting on the particle.

$$\vec{F}_{net} = \sum_n \vec{F}_n. \tag{5.7}$$

12. The "Law of Inertia" ("Newton's First Law") states that if the
net force $\vec{F}$ on a particle is zero, the particle retains its initial velocity
$\vec{v}_i$ relative to the remote stars (i.e. the particle retains its state of
rest or uniform straight line motion relative to the remote stars).

13. The "Main Law of Mechanics" ("Newton's Second Law") states that the acceleration of a particle relative to the remote stars is proportional to the net force acting on the particle

$$\vec{a} = \frac{\vec{F}_{net}}{m}.$$ (5.8)

The quantity m is the "mass" of the particle. It describes the resistance a particle offers to being accelerated (the greater the mass, the smaller the magnitude of the acceleration for the same magnitude of the net force).

14. The laws of mechanics are valid for all coordinate systems which move at constant velocity relative to the remote stars (such systems are called "inertial systems"). This statement is known as the "Principle of Relativity."

15. The relationship between the coordinates of two inertial systems S and S' is given by

$$\vec{r} = \vec{r}' + \vec{v}_o t,$$ (5.9)

where $\vec{v}_o$ is the velocity of the origin of the system S' relative to the origin of the system S. We suppose that the axes of S' are parallel to the axes of S and that time is measured from the moment when the origins of S and S' have the same position ($\vec{r} = \vec{r}'$ at $t = 0$). This transformation is known as the "Galilean Transformation."

16. The connection between the velocities in two inertial systems is given by

$$\vec{v} = \vec{v}' + \vec{v}_o.$$ (5.10)

The acceleration of a particle is the same in both inertial systems ($\vec{a} = \vec{a}'$).

17. When studying mechanics, we assume that particles do not have a net electric charge or a magnetic moment. In this case the only force between remote particles is the "gravitational force." However, when particles are in contact, they experience additional "contact forces."

18. "Newton's Law of Gravitation" states that the gravitational force exerted by particle 1 on particle 2 is given by

$$\vec{F}_{12} = -G\frac{m_1 m_2}{r_{12}^2}\vec{n}_{12},\tag{5.11}$$

where G is the universal gravitational constant, r_{12} is the distance between the two particles, m_1 and m_2 are their masses and $\vec{n}_{12}$ is the unit vector which points from particle 1 to particle 2. The negative sign shows that $\vec{F}_{12}$ is opposite to $\vec{n}_{12}$, i.e. particle 2 is attracted to particle 1.

19. A large solid rigid object can be considered as a system of many particles connected to each other. The position of such an object is described by the position $\vec{r}$ of its "center of mass"

$$\vec{r} = \frac{\sum m_n \vec{r}_n}{m} \approx \frac{\int \vec{r}\,'\,dm\,'}{m},\tag{5.12}$$

where m is the total mass of the object.

20. The center of mass of any system of particles (in particular, the center of mass of a solid object) moves like a single particle according to the main law of mechanics

$$\vec{F}_{net} = m\vec{a},\tag{5.13}$$

where $\vec{F}_{net}$ is the total net force on the system of particles, m is its total mass, and $\vec{a}$ is the acceleration of the center of mass.

21. The gravitational force between two uniform spherical objects is given by the Newton's Law of Gravitation (5.11), where r_{12} is now the distance between the centers of the spherical objects, and $\vec{n}_{12}$ is the unit vector that points from the center of object 1 to the center of object 2.

22. For "terrestrial phenomena" the characteristic time is often very much smaller than one day. In this case we can consider a coordinate system connected to the ground as an inertial system. For terrestrial phenomena, the magnitude of the gravitational force produced by

the Earth on an object of mass m can be approximated as

$$F_g = mg,\tag{5.14}$$

where g is the gravitational acceleration.

23. If a flat surface of one object is in contact with a flat surface of another object, the contact force exerted by object 1 on object 2 can be represented as a sum of two forces: the normal force $\vec{\mathbf{F}}_n$ (the force of repulsion, which is perpendicular to the surfaces), and a tangential friction force.

24. If object 2 is sliding relative to object 1, the tangential force on object 2 is called the "kinetic friction force" $\vec{\mathbf{F}}_k$. This force is opposite to the direction of the relative velocity $(\vec{\mathbf{v}}_2 - \vec{\mathbf{v}}_1)$. The magnitude of this force is related to the magnitude of the normal force by the approximate equation

$$F_k = \mu_k F_n,\tag{5.15}$$

where μ_k is called "the coefficient of kinetic friction."

25. If object 2 is not sliding relative to object 1 the tangential force, which prevents sliding, is called the "static friction force" $\vec{\mathbf{F}}_s$. The magnitude of F_s cannot be greater than the product $\mu_s F_n$ $(F_s \leq \mu_s F_n)$, where μ_s is called "the coefficient of static friction."

26. Let an object be attached to a rope (or string or wire). If the rope is taut, the force exerted by the rope on the object points in the direction of the rope. This force is often called the "tension force."

27. Let an object be attached to a spring, which can be compressed or stretched, for example, in the x direction. The force exerted by the spring on the object is given by

$$\vec{\mathbf{F}} = -\vec{\mathbf{i}}k_s(x - x_0),\tag{5.16}$$

where k_s is the spring (or force) constant of the spring, x is the position of the end of the spring, and x_0 is its equilibrium position. (We suppose that the other end of the spring is fixed at $x < x_0$.)

This force is called the "spring force" or the "elastic force." If $x > x_0$ the spring force points in towards the spring, if $x < x_0$ it points away from the spring.

28. If the net force $\vec{\mathbf{F}}_{net}$ on a particle is constant, the position of the particle is given by

$$\vec{\mathbf{r}} = \vec{\mathbf{r}}_i + \vec{\mathbf{v}}_i t + \frac{\vec{\mathbf{a}}t^2}{2}, \quad \vec{\mathbf{a}} = \frac{\vec{\mathbf{F}}_{net}}{m}, \qquad (5.17)$$

where $\vec{\mathbf{r}}_i$ and $\vec{\mathbf{v}}_i$ are the initial position and initial velocity of the particle (at time $t = 0$). The velocity $\vec{\mathbf{v}}$ of the particle is given by

$$\vec{\mathbf{v}} = \vec{\mathbf{v}_i} + \vec{\mathbf{a}}t . \qquad (5.18)$$

Two other expressions are useful when the acceleration is constant,

$$\vec{\mathbf{r}}_f = \frac{1}{2}(\vec{\mathbf{v}}_i + \vec{\mathbf{v}}_f)t_f , \qquad (5.19)$$

$$v_{\alpha f}^2 = v_{\alpha i}^2 + 2a_\alpha(\alpha_f - \alpha_i) ,$$

where α is x, or y, or z, and the index "f" means final.

29. The "momentum" of a particle $\vec{\mathbf{p}}$ is defined as the product of its mass and velocity

$$\vec{\mathbf{p}} = m\vec{\mathbf{v}} . \qquad (5.20)$$

The rate of change of the momentum is equal to the net force on the particle

$$\frac{d\vec{\mathbf{p}}}{dt} = \vec{\mathbf{F}}_{net} . \qquad (5.21)$$

The same equation is valid for a system of particles, where

$$\vec{\mathbf{p}} = \sum m_n \vec{\mathbf{v}}_n \qquad (5.22)$$

is the total momentum of the system, and $\vec{\mathbf{F}}_{net}$ is the total net force on the system.

30. If the net force on a system of particles equals zero, the momentum of the system does not change and so is conserved (the "Law

of Conservation of Momentum"). If any component of the net force equals zero, the same component of the momentum is conserved.

31. The impulse delivered by a force $\vec{\mathbf{F}}_k$ to a particle during the time interval $t_i < t < t_f$ is defined as

$$\vec{\mathbf{I}}_k = \int_{t_i}^{t_f} \vec{\mathbf{F}}_k \, dt \,. \tag{5.23}$$

The impulse delivered by a net force equals the change in the particle's momentum ("The Impulse–Momentum Theorem")

$$\vec{\mathbf{I}} = \vec{\mathbf{p}}_f - \vec{\mathbf{p}}_i \,. \tag{5.24}$$

32. The kinetic energy of a particle is defined as

$$K = \frac{mv^2}{2} \,. \tag{5.25}$$

The rate of change of the kinetic energy is given by

$$\frac{dK}{dt} = \vec{\mathbf{F}} \cdot \vec{\mathbf{v}} \,. \tag{5.26}$$

The vector dot product $\vec{\mathbf{F}} \cdot \vec{\mathbf{v}}$ is called the "power" of the force $\vec{\mathbf{F}}$.

33. The work done by a force $\vec{\mathbf{F}}_k$ on a particle moving from the point P_i to the point P_f is defined as

$$W = \int_{P_i}^{P_f} \vec{\mathbf{F}}_k \cdot d\vec{\mathbf{r}} \,. \tag{5.27}$$

The work done by the net force is equal to the change in the particle's kinetic energy ("Work–Kinetic Energy Theorem")

$$W = K_f - K_i \,. \tag{5.28}$$

34. The work done by the gravitational force on a particle of mass m near the surface of the Earth (terrestrial phenomenon) is given by

$$W_g = mg(y_i - y_f) \,, \tag{5.29}$$

where y_i and y_f are the initial and final coordinates of the particle, assuming that the y-axis points vertically upwards. The function

$$U_g(y) = mgy + \text{constant} \qquad (5.30)$$

is called the gravitational potential energy.

35. The kinetic energy of a system of particles is defined as the sum of the kinetic energies of all the particles

$$K = \sum K_n . \qquad (5.31)$$

The gravitational potential energy of a system of particles is the same as for a single particle, $U(y) = mgy + \text{constant}$, where m becomes the total mass of the system, and y becomes the y-coordinate of its center of mass.

36. The work done by a spring force on an object, which is in contact with the spring's free end and is moving along the spring axis (x-axis), is given by

$$W_s = \frac{1}{2}k_s(x_i^2 - x_f^2), \qquad (5.32)$$

where x_i and x_f are initial and final positions of the spring end. The function

$$U_s(x) = \frac{1}{2}k_s x^2 + \text{constant} \qquad (5.33)$$

is called the spring (or elastic) potential energy.

37. The mechanical energy E of a system of particles (or a simple solid object) is defined as the sum of its kinetic and potential energies

$$E = K + \sum U_n . \qquad (5.34)$$

If only gravitational and spring forces do work on the system of particles, then the mechanical energy of the system is conserved (the "Law of Conservation of Mechanical Energy"). The gravitational and spring forces are called "conservative forces." The work done on a particle by conservative forces depends only on the initial and final positions of the particle.

38. Let a particle move around a circle of radius r in circular motion. The acceleration of the particle has two components: the centripetal component, $\vec{a}_c$, which points towards the center of the circle, and the tangential component, $\vec{a}_t$, which is tangential to the circle. The magnitude of the centripetal component is given by

$$a_c = \frac{v^2}{r} \,, \tag{5.35}$$

where v is the speed of the particle and the tangential component of the acceleration is zero if v is constant.

39. Let a particle revolve about the z-axis, in the x–y plane. The infinitesimally small angular displacement $\mathbf{d}\vec{\theta}$ can be described by the vector

$$\mathbf{d}\vec{\theta} = \vec{\mathbf{k}}(d\theta) \,. \tag{5.36}$$

The angular velocity $\vec{\omega}$ is defined as

$$\vec{\omega} = \frac{d\vec{\theta}}{dt} \,, \tag{5.37}$$

and its magnitude is the angular speed ω.

The angular acceleration $\vec{\alpha}$ is

$$\vec{\alpha} = \frac{d\vec{\omega}}{dt} \,. \tag{5.38}$$

40. The connections between v, ω, a_t and α are given by

$$v = r\omega, \quad a_t = r\alpha \,. \tag{5.39}$$

41. Let a solid object rotate about a fixed axis (z-axis). The position of the rotating object can be described by the angle θ between the x-axis and the position vector for any point in the object revolving in the x–y plane (x–y plane "cuts" through the object). Angular position, displacement, velocity and acceleration for the rotating object are defined in the same way as for a particle.

42. The torque $\vec{\tau}_k$ produced by the force $\mathbf{F}_k$ is defined as the vector cross product:

$$\vec{\tau}_k = \mathbf{r}_k \times \mathbf{F}_k, \tag{5.40}$$

where $\mathbf{r}_k$ is the position vector of the point of application of the force.

43. The moment of inertia I of a particle revolving about the z-axis is defined as

$$I = m(x^2 + y^2). \tag{5.41}$$

The moment of inertia of a solid object rotating about the z-axis is

$$I = \int (x^2 + y^2)dm. \tag{5.42}$$

44. The "Law of Rotation" is similar to the Main Law of Mechanics and is also know as "Newton's Second Law for Rotation"

$$\vec{\alpha} = \frac{\vec{\tau}_{net}}{I}, \tag{5.43}$$

where $\vec{\tau}_{net}$ is the net torque (vector sum of all torques) acting on the object.

45. The angular momentum of a particle is defined as the cross product

$$\vec{\mathbf{L}} = \vec{\mathbf{r}} \times \vec{\mathbf{p}}. \tag{5.44}$$

For a solid object rotating about a fixed axis, the angular momentum is

$$\vec{\mathbf{L}} = I\vec{\omega}. \tag{5.45}$$

46. The rate of change of a particle's angular momentum is equal to the net torque on the particle

$$\frac{d\vec{\mathbf{L}}}{dt} = \vec{\tau}. \tag{5.46}$$

The same equation is valid for a system of particles, and in particular, for a solid object.

47. If the net torque on a system of particles equals zero, the angular momentum of the system does not change (the "Law of Conservation of Angular Momentum").

48. The rotational kinetic energy of a revolving particle can be written as

$$K = \frac{I\omega^2}{2} \,. \tag{5.47}$$

The same equation is valid for a rotating solid object.

49. The conditions for static equilibrium of an object are the following:

$$\vec{\tau} = 0, \quad \vec{F} = 0 \,, \tag{5.48}$$

assuming $\vec{v}_i = 0$ and $\vec{\omega}_i = 0$.

50. To describe celestial phenomena in the solar system , it is convenient to use a coordinate system fixed to the center of the Sun. It is a very good approximation to assume this is an inertial system. To describe the motion of a satellite about the Earth it is convenient to use a coordinate system fixed at the Earth's center. For the satellite's motion, this can be considered an inertial system.

51. Planets orbit the Sun moving along elliptical paths ("Kepler's First Law"). The ratio of the squared period T^2 to the cubed semi-major axis R_1^3 is the same for all the Sun's planets ("Kepler's Third Law").

52. If the net force on an object is the spring force $\vec{F} = -\vec{i}k_s(x - x_0)$, the motion of the object (rigorously, the motion of the end of the spring which is attached to the object) is given by

$$x = x_0 + A\cos(\omega t + \phi) \,. \tag{5.49}$$

This equation describes the harmonic oscillation of an object with amplitude, A, angular frequency, ω, and phase constant, ϕ. The frequency of the oscillation is given by

$$f = \frac{\omega}{2\pi},$$ (5.50)

and the period, T, is

$$T = \frac{1}{f} = \frac{2\pi}{\omega}.$$ (5.51)

53. A disturbance traveling through a substance without permanent displacements of the particles of the substance is called a traveling wave. If the disturbance can be described by a harmonic function (sine or cosine) the traveling disturbance is called a harmonic traveling wave.

54. The simplest example of a harmonic traveling wave is a wave on a long taut string. If one of the ends of the string starts to move in a simple harmonic oscillation, e.g. along the y-axis, the oscillation "travels" along the string in the x-direction, if the x-axis points along the string (see Fig. 7).

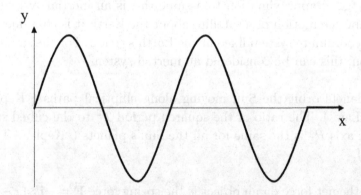

Fig. 7.   A harmonic traveling wave.

55. The y-coordinate of a point on a string in a harmonic traveling wave is given by

$$y(x,t) = A\sin(kx - \omega t + \phi), \quad 0 \le x \le \frac{\omega t}{k},$$ (5.52)

if the wave travels in the positive x-direction, or

$$y(x,t) = A\sin(kx + \omega t + \phi), \quad -\frac{\omega t}{k} \leq x \leq 0, \qquad (5.53)$$

if the wave travels in the negative x-direction. (We assume that the wave is produced at the origin beginning at time $t = 0$.)

56. The quantity k is called the wave number, ω is the angular frequency, ϕ is the phase constant, and A is the amplitude of the wave. The angle $(kx \mp \omega t + \phi)$ is called the phase of the wave. These parameters are common for all harmonic traveling waves. Points where the displacement from the equilibrium position is the largest are called the crests and troughs of the wave.

57. The frequency of a harmonic wave is defined as

$$f = \frac{\omega}{2\pi} . \qquad (5.54)$$

58. The period of a harmonic wave is

$$T = \frac{1}{f} = \frac{2\pi}{\omega} . \qquad (5.55)$$

59. The wavelength, λ, of a harmonic wave is the distance between two neighboring points with the same phase

$$\lambda = \frac{2\pi}{k} . \qquad (5.56)$$

60. The speed of a harmonic wave is

$$v = \frac{\omega}{k} = \frac{\lambda}{T} = \lambda f . \qquad (5.57)$$

All these relations are valid for all harmonic traveling waves.

61. The speed of a traveling wave in a string is given by

$$v = \sqrt{\frac{F_T}{\mu}} , \qquad (5.58)$$

where F_T is the tension force in the string and $\mu = m/L$ is the linear mass density of the string (of mass m and length L).

62. The power transmitted by a harmonic traveling wave in a string is given by

$$P = \frac{1}{2}\mu v(A\omega)^2 . \qquad (5.59)$$

63. The intensity of a sound wave, I_s is the power transmitted by the sound wave P_s per unit area

$$I_s = \frac{dP_s}{dA}, \qquad (5.60)$$

where the surface of area element dA is assumed to be perpendicular to the direction of propagation of the sound wave.

64. The level of a sound wave β is defined as

$$\beta = 10 \log \frac{I_s}{I_{so}}, \qquad (5.61)$$

where $I_{so} = 10^{-12}$ W/m^2 is conventionally called the "threshold of hearing."

65. When a source of a sound wave and (or) an observer who detects a sound wave move relative to the air, the frequency f' of the detected wave is given by (the "Doppler Effect")

$$f' = \frac{v \pm v_o}{v \mp v_s} f , \qquad (5.62)$$

where f is the frequency of the stationary source detected by a stationary observer, v is the speed of sound, v_o is the speed of the observer, v_s is the speed of the source. The upper signs (frequency increase) correspond to the source moving toward the observer and/or the observer moving toward the source, while the lower signs correspond to the opposite cases.

66. A harmonic standing wave in a string is described by the equation for the y-coordinate of a point in the string

$$y = A \sin (kx + \phi_1) \cos (\omega t + \phi_2) . \qquad (5.63)$$

67. The amplitude of a harmonic standing wave $A \sin (kx + \phi_1)$ is a harmonic function of x. The points which do not oscillate are called "nodes," the points which oscillate with the maximum amplitude A (amplitude of the standing wave) are called "antinodes." All points between two neighboring nodes oscillate with the same phase $(\omega t + \phi_2)$ or $(\omega t + \phi_2 + \pi)$.

68. The wavelength of a harmonic standing wave, $\lambda = 2\pi/k$, is the distance between two neighboring points of the same amplitude and phase. The standing wave does not have a speed because the oscillations "fill" the whole string and the nodes and antinodes do not move along the string. The period of the wave is $T = 2\pi/\omega$, and the frequency is $f = 1/T = \omega/2\pi$. The possible frequencies of the harmonic standing waves take on discrete values f_n. The minimum frequency is called the fundamental frequency, and the corresponding standing wave is called the fundamental mode. All frequencies f_n are called natural frequencies. All these properties and terms are common for all harmonic standing waves.

69. For a string fixed at both ends the fundamental frequency is

$$f_1 = \frac{v}{2L}, \tag{5.64}$$

where L is the length of the string, and v is the speed of the harmonic traveling wave in the string. All natural frequencies are given by

$$f_n = nf_1, \quad \text{where} \quad n = 1, 2, 3, \ldots \tag{5.65}$$

The corresponding natural wavelengths are

$$\lambda_n = \frac{v}{f_n} = \frac{2L}{n}, \quad \text{where} \quad n = 1, 2, 3, \ldots \tag{5.66}$$

70. For a string fixed at one end, the natural frequencies are given by

$$f_n = nf_1, \quad n = 1, 3, 5, \ldots, \tag{5.67}$$

where the fundamental frequency is

$$f_1 = \frac{v}{4L}.$$ (5.68)

The corresponding natural wavelengths are

$$\lambda_n = \frac{v}{f_n} = \frac{4L}{n}, \quad n = 1, 3, 5, \ldots$$ (5.69)

5.2. Problems

1. The troposphere (where the weather takes place) is the lower part of the Earth's atmosphere, extending for approximately 10 km above the ground. Estimate the relative change in the gravitational acceleration from the bottom to the top of the troposphere (in percent).

2. The presence of bacteria at a height of about 40 km above the ground was discovered in 2001. Scientists concluded that the bacteria arrive from outer space because nothing can transport them from the ground to such a great height (this conclusion is not universally accepted). Estimate the gravitational force on a bacterium at a height of 40 km, assuming its mass to be of the order of $1fg$.

3. The Voyager-2 probe was launched in 1977. It passed near Jupiter, Saturn, Uranus, and Neptune. It has now moved out of the solar system.

a) The speed of the Voyager-2 spacecraft is about 38,000 mph. Find its speed in SI units.

b) Assuming a constant speed, estimate the distance traveled by Voyager-2 in one year.

4. The Cosmos-1 project, privately funded by the Planetary Society, is designed to use the pressure of the Sun's radiation to move a solar sail spacecraft like an ordinary sail uses the pressure of the wind.

a) The area of the Cosmos-1 sail is about 720 square yards, and the weight of the spacecraft is about 88 pounds. Find the area and mass in SI units.

b) What must be the magnitude of the force exerted by the solar light on the sail to balance the gravitational force on the spacecraft exerted by the Earth at a distance of 100 Mm from the center of the Earth?

c) Let the solar sail spacecraft move away from the Earth and the Sun. At what distance from the center of the Earth will the gravitational force on the spacecraft exerted by the Earth be equal to the gravitational force exerted by the Sun? (Assume the Sun, the Earth and the spacecraft are on the same line, in this order.)

5. It is estimated that the interstellar solar sail probe can reach a speed of 380,000 mph, 10 times greater than Voyager-2 (see Problem 3).

a) Estimate the distance the interstellar probe can cover in one year at this speed.

b) Estimate the momentum and kinetic energy of the interstellar probe, assuming its mass is 40 kg.

6. The Galileo mission, launched in 1989, consisted of a 2400 kg orbiter and a 300 kg probe. The spacecraft became the first to orbit Jupiter. It traveled first toward the Sun, encountered Venus and then the Earth twice, so as to gain enough speed to reach Jupiter. Consider an encounter with the Earth at a distance of about 1 Mm from the ground.

a) Find the magnitude of the gravitational force exerted by the Earth on the spacecraft.

b) Estimate the magnitude of the acceleration of the spacecraft.

7. The Genesis spacecraft collected atoms from the solar wind and returned them to the Earth in 2004. The distance covered in this round trip mission was about 20 million miles and the trip took about 3 years. Estimate the average speed of the Genesis craft in SI units.

8. The Genesis spacecraft collected solar atoms in a region which is at a distance about 92 million miles from the center of the Sun and 1 million miles from the center of the Earth.

Estimate the ratio of the gravitational force exerted by the Sun on the spacecraft to the gravitational force exerted by the Earth in that region. (Consider only the magnitudes of the forces.)

9. In Problem 8, assume that a gold atom in the solar wind moving at 10^6 mph relative to the spacecraft was captured by a Genesis collector.

a) Find the magnitude of the impulse delivered by the atom to the spacecraft.

b) Is the total momentum of the spacecraft and the atom conserved in the collision?

c) Find the kinetic energy of the gold atom relative to the spacecraft before the impact. Is the total kinetic energy of the spacecraft and the atom conserved in the impact?

10. Assume that an iron atom in the solar wind travels at 10^6 mph relative to the center of the Earth.

a) If it moves along a line which passes through the center of the Earth, find the magnitude of its angular momentum relative to the center of the Earth.

b) Solve the same problem if it moves along a line which is tangential to the Earth's surface.

11. The Cassini spacecraft, launched in 1997, entered an orbit around Saturn in 2004. Estimate the gravitational force between the Sun and the 2000 kg orbiter traveling around Saturn. Assume the distance between Saturn and the Sun to be 9.5 astronomical units (AU).

12. In 2005, spacecraft Cassini delivered a probe called Huygens to Saturn's moon Titan. Estimate the gravitational acceleration near Titan's surface. Assume the radius of Titan to be 2.6 Mm and its mass to be 1.35×10^{23} kg.

13. The 12,600 pound Cassini spacecraft passed Jupiter at a distance of 6 million miles from its center in December 2000. Find the gravitational force between the spacecraft and Jupiter at this distance.

14. Images taken by Cassini confirmed the existence of long-lived Jovian storms, which can last many years and probably several centuries!

a) Assuming 300 mph winds in the Jovian atmosphere, find the momentum (magnitude) and kinetic energy of an ammonia molecule (NH_3) moving with the speed of the wind.

b) Estimate the magnitude of the angular momentum of the molecule relative to the center of Jupiter. Assume the radius of Jupiter to be 7.18×10^7 m.

15. A comet nucleus is a small celestial body consisting of ice and dust. Assuming that a comet is a pure ice ball of 3 km diameter, estimate its mass. Take the density of ice to be 917 kg/m^3.

16. As a comet approaches the Sun, it releases gas and dust under the action of the Sun's radiation . This material forms a diffuse cloud of gas and dust surrounding the nucleus of the comet called the "coma." Assuming that the mass of the coma is 10% of the total mass estimated in problem 15, and it has a spherical form 100 megameters in diameter, estimate the average density of the coma.

17. Typically a comet orbits the Sun in a highly elliptical orbit. "Long-period comets" have periods of more than 200 years. Some of them have periods of up to 10^7 years! Assume that a comet has a perihelion distance (the closest distance to the Sun) of 3 astronomical units (AU) and aphelion distance (the maximum distance from the Sun) 10^5 AU.

a) As a first approximation, consider its motion to be a straight line oscillation and estimate the average speed of the comet.

b) Find the ratio of gravitational forces (magnitudes) between the comet and the Sun at aphelion and perihelion.

18. In July 1994 the comet "Shoemaker–Levy" collided with Jupiter. It is estimated that long before the collision the comet nucleus split into about 20 fragments. Assume that a spherical pure ice fragment 1 km in diameter crashed into Jupiter at 60 km/s. Estimate the kinetic energy and the magnitude of the fragment's momentum just

before the collision. Estimate the magnitude of impulse delivered by the fragment to Jupiter. Take the density of ice to be 917 kg/m^3.

19. It is estimated that the nucleus of the Shoemaker–Levy comet split apart under the action of "tidal forces" (i.e. the nonuniformity of the gravitational force across a body) at a distance of about 20 Mm from Jupiter. Estimate the difference between the magnitude of the gravitational force produced by Jupiter on two particles of the same mass m on the surface of the comet but assuming that one particle is 5 km closer to Jupiter's center than the other one.

20. About 10 times per year the Earth passes through swarms of cosmic dust particles, causing "meteor showers." A piece of this cosmic dust is called a "meteoroid." The typical size of a meteoroid is 1 cm, and its mass is about 1 g. Meteoroids burn up in the Earth's atmosphere producing meteors ("shooting stars"). A typical meteoroid begins to shine at a height of about 90 km above the ground.

    a) Estimate the density of a typical meteoroid.

    b) Estimate the acceleration of a meteoroid at a height of 90 km.

    c) Assume a meteoroid "collides" with the Earth's atmosphere at a speed of 50 km/s. Estimate the magnitude of the momentum and the kinetic energy of such a typical meteoroid.

21. In 2001, the solar-powered unmanned aircraft Helios reached an altitude of about 9.69×10^4 ft. The previous altitude record for a propeller-drive plane was about 8.59×10^4 ft. Find the difference between these two altitudes in meters.

22. The 62,000 solar cells on the Helios aircraft generated about 40 kW of power to drive its 14 propellers. How much energy was needed to drive one propeller for an hour?

23. Assume the Helios aircraft is flying so as to remain essentially stationary above the same point on the equator, at a height of 9.69×10^4 ft above the ground.

    a) Find the speed of the aircraft relative to the center of the Earth.

    b) Find the magnitude of the acceleration of the aircraft relative to the center of the Earth.

c) Find the ratio of the magnitude of the gravitational force exerted on the aircraft by the Earth to the total force exerted on it by the air (magnitudes). (Note that the net force on Helios is sum of the gravitational force and the total force exerted by the air.)

d) Find the ratio of the magnitudes of the aircraft's angular momentum with respect to the center of the Earth and its tangential linear momentum.

24. Assume the Helios aircraft traveled at 150 mph relative to the ground. Find the ratio of its translational kinetic energy to the magnitude of its tangential linear momentum (in m/s) relative to the ground.

25. In experiments with a high energy beam of lead ions colliding with stationary lead ions, the nuclear matter density was estimated to be 20 times that of normal nuclear matter (e.g., of a proton). Estimate the density achieved in such experiments (in kg/m^3). Use 1.5 femtometers as the radius of a proton.

26. The normal Atlantic hurricane season lasts from June until December. Typically it brings two or three major hurricanes with wind speeds greater than 110 mph. What is the kinetic energy (in joules) of 1 m^3 of air, if it moves at 120 mph? Assume the density of air to be 1.3 kg/m^3.

27. In problem 26 what is the magnitude of the momentum of 1 m^3 of air moving at 120 mph(in Ns)?

28. A major hurricane "sustains" winds greater than 110 mph. Find the kinetic energy in joules of a nitrogen molecule moving with the speed of the wind, 110 mph.

29. The NEAR spacecraft softly landed on the asteroid Eros. The landing speed of the 1,100 pound NEAR did not exceed 1.8 m/s. Assuming this value for the speed estimate:

a) The momentum (magnitude) and kinetic energy of the spacecraft.

b) The impulse (magnitude) delivered by Eros' surface on the spacecraft.

30. Assume that the International Space Station (ISS) orbits the Earth at an altitude of 230 miles above the ground. Assuming a circular orbit find the speed, angular speed, acceleration (magnitude), and orbital period of ISS.

31. If the distance between the orbiting ISS and the center of the Earth increases 9 times, how would it affect the speed, angular speed, acceleration, and orbital period of the ISS?

32. If the distance between the orbiting ISS and the center of the Earth increases by a factor of 9, how would it affect the momentum, angular momentum, and kinetic energy of ISS?

33. NASA's laboratory on the ISS is called "Destiny." It is 8.4 m long, and has a diameter of 4.2 m. If the weight of Destiny is 30,000 pounds, what is its average density (in kg/m^3)?

34. Taking data from problems 30 and 33 find:

a) The gravitational force (magnitude) exerted by the Earth on Destiny.

b) The momentum, angular momentum (magnitudes) and kinetic energy of Destiny.

35. Estimate the ratio of the gravitational forces exerted by the Earth on the ISS to the average force exerted by the Sun on the ISS (magnitudes). Take the altitude of the ISS to be 230 miles.

36. The Boeing Delta 4 launch vehicle can launch payloads of mass about 13,000 kg. Assume a Delta 4 delivers a 13 Mg payload to a geosynchronous orbit (the period of a geosynchronous orbit is one day). Find the final momentum, angular momentum and kinetic energy of the payload in a geosynchronous orbit.

37. In 2000 astronomers discovered the small planet ("Plutino") 2000EB173 between the orbits of Neptune and Pluto. Plutino, which is about 400 miles in diameter, orbits the Sun with a period estimated to be 240 years. Estimate the semimajor axis of Plutino's orbit (in astronomical units).

38. The most distant galaxies scientists can now see are about 10 billion light-years (10^{10} ly) away. Estimate the volume of the visible universe (in m^3).

39. A typical "large" asteroid, with a size of 100 km and density about 2.5 g/cm^3, rotates at 5 revolutions per day. Assuming that the asteroid has an approximately spherical shape, estimate (in a coordinate system fixed at the asteroid's center):
    a) the angular speed of the asteroid (in rad/s);
    b) the speed of a point near the asteroid's equator;
    c) the magnitude of the angular momentum (the moment of inertia for a uniform sphere is 2 $mr^2/5$);
    d) the rotational kinetic energy.

40. Solve problem 39:
    a) for a very large asteroid of size 800 km, that rotates at 2 revolutions per day;
    b) for a very small asteroid of size 50 m, that rotates at 500 revolutions per day.

41. The asteroid Castalia has a so-called Earth crossing orbit. In 1989 it was detected by radar at a distance of about 11 lunar distances from the Earth. Castalia's dimensions have been found to be about $1.8 \times 0.8 \times 0.8$ km, with a density of 2.1 g/cm^3 and a rotational period of 4 hours. Based on these data estimate:
    a) Castalia's mass (assume it has a cylindrical shape);
    b) the gravitational force between Castalia and the Earth at 11 lunar distances from the Earth;
    c) the angular speed of Castalia's rotation about its center.

42. On its way to Jupiter, the Galileo probe passed through the

main asteroid belt (between the orbits of Mars and Jupiter). Unexpectedly, Galileo detected an asteroid with a satellite, "Dactyl," which orbits the asteroid "Ida." The dimensions of Ida are approximately $60 \times 20 \times 20$ km and its density is 2.5 g/cm^3. The size of Dactyl is about 1.4 km. Assuming the same density for Ida and Dactyl find the ratio of the gravitational forces exerted by the Sun on the two celestial bodies. Assume a cylindrical shape for Ida and a spherical shape for Dactyl.

43. The dark space in our Galaxy (the "Milky Way") is associated with "clouds" of interstellar dust. A typical particle of dust, which blocks visible light, is an elongated grain with dimensions about $0.4 \times 0.2 \times 0.2$ μm. It consists of a rocky core and an ice mantle and has a mass of about 10^{-14} g. Find the average density of a grain assuming it has a cylindrical shape.

44. It is estimated that on average every 10^6 m^3 of interstellar space contains one dust particle. Using data from problem 43, estimate:
  a) The average distance between nearest neighbor dust grains.
  b) What fraction of interstellar space is occupied by dust particles?
  c) The average density of matter in interstellar space.
  d) The average gravitational force between two nearest-neighbor dust particles.

45. The cantilever used in magnetic resonance force microscopy is a tiny "beam," which is fixed at one end and free to vibrate at the other end. The free end of the cantilever oscillates approximately in simple harmonic motion like a small block attached to a massless spring. It can be described by an effective mass and an effective spring constant.

Consider an ultra-thin silicon cantilever (thickness about 60 nm, length about 200 μm). If the frequency of the cantilever vibrations is 1.7 kHz, and the effective spring constant is 6.5×10^{-6} N/m, find the angular frequency, the period and the effective mass of the cantilever.

46. Let the cantilever tip in problem 45 oscillate along the x-axis

near the origin. At time $t = 0$ let the cantilever tip be at the end-point of its motion at $x = 40$ nm.

   a) Find the position, velocity and acceleration of the cantilever tip at time $t = 1$ ms.

   b) When does the cantilever tip pass through the origin for the first time, and for the second time?

   c) When does the cantilever tip pass through the point $x = 20$ nm for the first time, and for the second time?

47. Let the cantilever tip in problem 45 oscillate with an amplitude of 30 nm.

   a) What is the maximum speed of the cantilever tip?

   b) What is the speed of the tip at a distance of 10 nm from the equilibrium position?

   c) What are the maximum magnitudes of the acceleration and the net effective force on the cantilever tip?

   d) What are the magnitudes of the acceleration and the net force at a distance of 10 nm from the equilibrium position?

48. Assume that the magnetic moment of a single proton produces a magnetic force of magnitude 6×10^{-19} N on a small magnet attached to a cantilever tip. If the force acts in the direction of the maximum possible displacement of the tip, estimate the magnitude of its equilibrium displacement. Take data from problem 45.

49. It is estimated that a maglev (a train which is levitated due to the action of magnetic forces) could achieve a speed 500 km/h. If the acceleration of the train is 1/5 of the gravitational acceleration, how long does it take to achieve this speed starting from rest?

50. In problem 49, find the distance traveled by the train before it achieves a speed of 500 km/h.

51. In problem 49, find the magnetic force (magnitude) needed to levitate a 55, 000 kg train car.

52. In problem 49, assume that the power dissipated by air resistance at a speed of 500 km/h is about 10 MW. Estimate the force of

air resistance on the train.

53. In problem 49, find the kinetic energy and momentum (magnitude) of a 55,000 kg train car moving at a speed of 500 km/h.

54. In problem 49, estimate the rate of change of momentum (magnitude) for a 55,000 kg train car, as it accelerates. What is the connection between this rate and force?

55. Studies show that the cost of launching a satellite could be significantly reduced if the rocket is accelerated up a long incline before the rocket engines are fired up. The speed of the rocket must be about 950 km/h before its engines are fired up, in order to obtain the desired savings. Assume that a "launcher" with the rocket and payload has a total mass of 2000 kg. Let the launcher be accelerated along a 30° incline (in mountains), and the length of the track be 1 km.
    a) What acceleration is needed to achieve a speed of 950 km/h?
    b) What driving (external) force (parallel to the incline) is needed to provide this acceleration? (Ignore friction and air resistance)

56. In problem 55, estimate how long it would take to achieve a speed of 950 km/h.

57. In problem 55, estimate the speed of the launcher:
    a) after the first second of acceleration;
    b) after it has covered the first 100 m of the track.

58. In problem 55, assume that the energy dissipated by resistive forces (air resistance, any kind of friction) is 100 MJ.
    a) How much energy is needed to accelerate the launcher to 950 km/h?
    b) What must be the average power supplied by the driving force?

59. In problem 55, assume that the driving force is shut off after the first second of acceleration.
    a) Find the distance traveled by the launcher (including the dis-

tance traveled during the first second) before the launcher stops momentarily. (Ignore resistive forces)

b) How long does it take to cover this distance?

60. In problem 55, assume that the driving force is shut down after the first 10 m of acceleration. Next, assume that the "resistive force" F_r in this case is similar to the kinetic friction force and can be described by the expression $F_r = \mu F_n$, where F_n is the normal force and $\mu = 0.1$. The launcher continues to move up the incline, then it stops momentarily, and then it slides all the way back down to the bottom of the incline.

a) From the moment when the driving force is shut down, how long does it take for the launcher to return to the bottom of the incline?

b) Find the distance traveled by the launcher during that time.

c) Find the magnitude of the launcher's displacement during that time.

61. Consider a tsunami generated in the deep ocean and moving toward a shore 4000 km away. Let the speed of the wave be 150 m/s and its wavelength be 700 km.

a) How long does it take tsunami to reach the shore?

b) Find the period of the wave.

c) Find the frequency of the wave.

62. Consider the front wall of a tsunami, of height 20 m and area 20 km $\times$ 100 m, hitting the shore at 10 m/s. Take the density of seawater to be 1025 kg/m^3.

a) Find the kinetic energy of the front wall.

b) Find the gravitational potential energy of the front wall (take the gravitational potential energy to be zero on the ground)

c) Find the momentum (magnitude) of the front wall.

63. In problem 61, imagine that a detector located near the ocean bottom has detected the change in pressure associated with the tsunami. This information is transmitted with a sound wave to a buoy on the ocean surface. Then the information is transmitted with an electro-

magnetic wave to a satellite and from the satellite to the shore. Let the detector be 2 km beneath the buoy, the distance between the satellite and the buoy be 1000 km, and the distance between the satellite and the shore station be 5000 km. How long does it take to transfer information from the detector to the shore station? Take the speed of the electromagnetic wave to be the speed of light 3×10^8 m/s, and the speed of the sound wave in the water to be 1.4 km/s.

64. It has been found that low-frequency sound (below 1 kHz) does not attenuate quickly in water. A modern low-frequency active sonar (LFA) emits sound pulses in the frequency range 100–500 Hz. Assume that the frequency of a particular pulse is 200 Hz, and its duration is 10 s. Taking the speed of sound in the water as 1.4 km/s, estimate:
    a) the wavelength of the sound wave;
    b) the period of the sound wave;
    c) the length of the sound pulse in the water;
    d) the number of wavelengths in the sound pulse.

65. A tau lepton τ^- decays into a muon μ^-, a muon antineutrino $\bar{\nu}_\mu$, and a tau neutrino ν_τ. The angular momentum components along the x-axis are the following:

$$L_x = \frac{\hbar}{2}, \quad \text{for } \tau^-, \tag{5.70}$$

$$L_x = \frac{\hbar}{2}, \quad \text{for } \mu^-, \tag{5.71}$$

$$L_x = \frac{\hbar}{2}, \quad \text{for } \nu_\tau, \tag{5.72}$$

where $\hbar = h/2\pi$, and h is Planck's constant. What is the value of L_x for the muon antineutrino $\bar{\nu}_\mu$?

66. In problem 65 let the momentum of the muon $\vec{p}(\mu^-)$ be equal to momentum of the tau lepton $\vec{p}(\tau^-)$. What then is the relationship between the momenta of the two neutrinos?

67. 3.67 ms after its generation, a muon neutrino of a certain energy transforms to a tau neutrino with probability close to 100%. At what distance from the accelerator should you place a tau neutrino detector in order to detect the neutrino? The speed of the neutrino is close to the speed of light.

68. A Higgs boson moving with a speed $v = 10$ Mm/s, (which is much smaller than the speed of light c), decays into two photons with the same magnitude of momentum, approximately, $p = mc/2$, where $m = 125$ GeV/c^2 is the rest mass of the Higgs boson. Find the direction of motion of either photon with respect to the direction of motion of the Higgs boson.

69. The 2004 Indian Ocean tsunami was triggered by an earthquake. If the energy of the earthquake is estimated to be 10^{18} J and its duration was about 9 minutes estimate the power of the earthquake.

70. The 2004 Indian Ocean tsunami was detected as far as 8.5 Mm away from the epicenter, 16 hours after the earthquake. Estimate the average speed of the tsunami wave.

71. The 2004 Indian Ocean tsunami had an initial speed 800 km/h, and its energy was estimated as 2×10^{16} J. If a tsunami were instead triggered by an asteroid hitting the Earth at 800 km/h estimate the minimum mass of the asteroid.

72. Consider a binary star system, which may evolve into a type *Ia* supernova. Assume the first star is a White Dwarf of mass $1.3M_o$ (M_o is the mass of the Sun). The second star is an ordinary star of mass $1.1M_o$. Assume that the distance between the two stars is 18 AU (astronomical units), and they move in circular orbits.

a) Find the distance between the White Dwarf and the center of mass of the binary system,

b) Find the speed of the stars relative to their center of mass,

c) Find the angular momentum of the binary system relative to the center of mass,

d) Find the kinetic energy of the binary system relative to the center of mass.

73. One of the most unusual binary systems consists of a Black Hole and a small star orbiting the common center of mass. The distance between the two objects was found to be about 1 Gm, i.e. much less than the Earth–Sun distance. The mass of the Black hole was estimated to be about $4M_o$, and the mass of its companion about $0.2M_o$. Solve Problem 72 for this binary system.

74. Columbus, the European research laboratory on the International Space Station (ISS), has a mass of 12800 kg (including its contents). Estimate the energy required to deliver Columbus to the ISS. Assume that ISS moves in a circular orbit 230 miles above the ground. Ignore the daily rotation of the Earth.

75. The Planck space observatory, of mass 1950 kg, orbited the Sun in the "shadow" of the Earth, at a distance of 1.5 Gm from the Earth. It traveled "behind" the Earth near the straight line that passes through the centers of the Sun and of the Earth. Find the gravitational force
    a) between the Planck space observatory and the Earth,
    b) between the Planck space observatory and the Sun.

76. The six-wheeled, 900-kg, rover Curiosity was delivered to the surface of Mars in 2012. If the rover travels with its maximum speed of 90 m/hr estimate its maximum translational kinetic energy relative to the Martian surface.

77. In Problem 76, the diameter of the rover's wheel is 50 cm. Find the maximum angular speed of a wheel.

78. When the Mars Science Laboratory (MSL) entered the Martian atmosphere, its initial deceleration was about 15g. If the rate of change of its deceleration was constant and equal to 15g/2, estimate how long would it take to reduce the speed of the MSL from its initial value of 5.8 km/s to zero.

79. The first interplanetary solar sail IKAROS, accelerated by the pressure of the solar radiation, was launched by the Japan Aerospace Exploration Agency in May, 2010. In December of the same year the 315-kg sail was at a distance of 80800 km from the center of Venus. Estimate the gravitational force between the solar sail and Venus at that distance.

80. Gamma-ray bursts (GRB), detected by orbiting telescopes, are typically caused by the explosion of distant massive stars that result in the creation of Black Holes. One of the closest GRB sources was detected at a distance about five billion light-years away. Photons from this source had energies up to 95 GeV, which corresponds to a momentum of about 5×10^{-17} kg·m/s. Assume that such a photon is absorbed by a stationary 1-mg object. What will be the speed of the object after the absorption?

81. In accelerators, particles gain energy in two stages. First, their speed increases to nearly the speed of light. Then, the energy is boosted without much increase in speed. Assume that an electron with a speed close to the speed of light enters a "mini-accelerator," which boosts its energy at a rate of 500 MeV/m. Estimate the time it takes to increase the electron energy by 50 GeV.

82. Assume that the brightness of a planet and its star are equal at an infrared wavelength of 25 μm. If the planet orbits the star at a distance of 10 AU and the system is 10 light years away, what must the angular resolution of a telescope be to resolve the objects under ideal conditions?

83. Sirius is the brightest star in the night sky, with mass equal to two Solar masses, located at a distance of 8.6 light years from the Sun. Calculate the position of the center of mass of the Sun–Sirius system with respect to the center of the Sun.

84. Calculate the position of the barycenter (center of mass) of the solar system with respect to the center of the Sun, assuming the dominant other mass in the system is Jupiter.

85. Show that the depth of the minimum in the brightness curve for a transiting exoplanet is $(r_p/R_s)^2$ where r_p is the radius of the planet and R_s is the radius of the star.

86. What is the probability that a planet orbiting a Sun-like star at a distance of 1 AU will produce an observable transit event? The radius of the Sun, R_o, is 696 Mm.

87. In August 2014, after a 10-year trip, the Rosetta spacecraft, launched by European Space Agency(ESA), approached the comet 67P/Churyumov–Gerasimenko. At that time the comet was at a distance of about 500 Gm from the Earth. Estimate the magnitude of the average velocity of Rosetta.

88. In November 2014, the Philae probe, delivered by the Rosetta spacecraft, landed on the comet 67P. The weight of the Philae on the surface of the Earth was about 1 kN, while on the surface of the comet it was only 10 mN. Assuming the comet is a solid sphere of 4 km in diameter, estimate the mass of the comet in terms of the mass of the Earth M_e.

Chapter 6

Electricity and Magnetism

6.1. Review

1. An atom consists of a nucleus and electrons, and the nucleus consists of protons and neutrons. The interaction which is responsible for the attraction between the nucleus and electrons is the electric interaction, which is a special case of the electromagnetic interaction.

2. To describe the electric interaction, one assigns to an electron a negative electric charge of $-e$, to a proton a positive electric charge of e and to a neutron zero electric charge.

3. If the difference between the numbers of protons N_p and electrons N_e in a "common" macroscopic object is non zero, one says that the macroscopic object is electrically charged, i.e. it has a macroscopic electric charge

$$q = e(N_p - N_e). \tag{6.1}$$

The macroscopic electric charge can be positive (if $N_p > N_e$) or negative (if $N_p < N_e$).

4. "Coulomb's Law" states that the electric force exerted by electrically charged particle 1 on electrically charged particle 2 is given by

$$\vec{\mathbf{F}}_{12} = k_c \frac{q_1 q_2}{r_{12}^2} \vec{\mathbf{n}}_{12}, \tag{6.2}$$

where k_c is Coulomb's constant, r_{12} is the distance between the two charged particles, q_1 and q_2 are their electric charges and $\vec{\mathbf{n}}_{12}$ is the unit vector which points from particle 1 to particle 2. If the electric charges q_1 and q_2 have the same sign they repel each other, otherwise they attract each other.

5. The electric potential energy for two charged particles is given by

$$U = k_c \frac{q_1 q_2}{r_{12}} + \text{constant}. \tag{6.3}$$

6. The electric field produced by charged particle 1 at an arbitrary point P is given by

$$\vec{\mathbf{E}}_P = k_c \frac{q_1}{r_{1P}^2} \vec{\mathbf{n}}_{1P}, \tag{6.4}$$

where r_{1P} is the distance between charged particle 1 and the point P, and $\vec{\mathbf{n}}_{1P}$ is the unit vector which points from particle 1 to the point P. The electric field of a positive charge points out of the charge and the electric field of a negative charge points into the charge.

7. The electric field produced by a system of particles at the point P is the vector sum of all the electric fields produced by each charged particle (the "superposition principle"):

$$\vec{\mathbf{E}}_P = \sum_n \vec{\mathbf{E}}_{nP} = k_c \sum_n \frac{q_n}{r_{nP}^2} \vec{\mathbf{n}}_{nP}. \tag{6.5}$$

If an additional particle of charge q is placed at the point P, the electric force exerted by all the other charged particles on it is given by

$$\vec{\mathbf{F}}_P = q \vec{\mathbf{E}}_P. \tag{6.6}$$

For a positively charged particle, the direction of the electric force $\vec{\mathbf{F}}_P$ is the same as the direction of the electric field $\vec{\mathbf{E}}_P$. For a negatively charged particle, $\vec{\mathbf{F}}_P$ is directed opposite to $\vec{\mathbf{E}}_P$.

8. The electric flux through a surface S is defined as

$$\Phi_e = \int_S \vec{\mathbf{E}} \cdot d\vec{\mathbf{A}}. \tag{6.7}$$

The vector $d\vec{\mathbf{A}}$ is perpendicular to the surface S at any point on the surface, and its magnitude dA is equal to an infinitesimally small area around that point. For a closed surface, $d\vec{\mathbf{A}}$ points out of the enclosed volume, and for an open surface one can choose either of the two possible directions.

9. The electric flux through a closed surface in empty space is related to the net charge inside the surface by "Gauss's Law":

$$\Phi_e = \sum_n \frac{q_n}{\epsilon_o}, \qquad (6.8)$$

where $\epsilon_o = 1/4\pi k_c$ is called the permittivity of free space.

10. The electric field can be visualized using electric field lines. For an electric field produced by electrically charged particles, electric field lines start from positively charged particles or from infinity, and end on negatively charged particles or at infinity. The density of the electric field lines is proportional to the magnitude of the electric field. At any point on an electric field line, the electric field is tangential to the line at that point.

11. Let a system of fixed charged particles produce an electric field $\vec{\mathbf{E}} = \vec{\mathbf{E}}(\vec{r})$. If an additional charged particle moves from the point P_i to the point P_f, the work done by the electric force on the particle is given by

$$W = +q \int_{P_i}^{P_f} \vec{\mathbf{E}} \cdot d\vec{\mathbf{r}}, \qquad (6.9)$$

where q is the charge of the additional particle.

12. The quantity

$$V_P = -\int_{P_o}^{P} \vec{\mathbf{E}} \cdot d\vec{\mathbf{r}} \qquad (6.10)$$

is called the electric potential at point P relative to point P_o, where P_o is an arbitrarily chosen point. The electric potential energy of an additional charged particle of charge q placed at the point P is given by

$$U_P = qV_P + \text{constant}. \tag{6.11}$$

13. The electric potential produced by charged particle 1 at an arbitrary point P relative to "infinity" is given by

$$V_P = k_c \frac{q_1}{r_{1P}}. \tag{6.12}$$

The electric potential produced by a system of charged particles equals the sum of the individual electric potentials:

$$V_P = \sum_n V_{nP} = k_c \sum_n \frac{q_n}{r_{nP}}. \tag{6.13}$$

14. A conventional conductor contains a huge number of charged particles, which can move inside the conductor under the action of a macroscopic electric field. It is customary to consider free (mobile) positively charged particles inside the conductor, while typically the free particles are actually negatively charged electrons. For example, in copper there are more than 10^{22} mobile electrons per cubic centimeter (cm^{-3}). The results of the analysis do not depend on the sign of free charges.

15. Under conditions of electrostatic equilibrium, the macroscopic electric field inside a good conductor is zero, and all points in the conductor have the same electric potential. If a conductor has a macroscopic electric charge this charge is distributed on the conductor's surface. The electric field near the conductor's surface is perpendicular to the surface and its magnitude is

$$E = \frac{\sigma}{\epsilon_o}, \tag{6.14}$$

where σ is the local surface electric charge density.

16. The simplest capacitor consists of two conductors separated by a dielectric (insulator). The conductors carry the same magnitude but opposite sign electric charges. The charge on the capacitor Q is defined as the magnitude of the charge on either conductor, and the

voltage on the capacitor V is the absolute value of electric potential difference between the two conductors. The value of the stored charge Q is proportional to the voltage V:

$$Q = CV . \tag{6.15}$$

The coefficient C is called the capacitance of the capacitor.

17. For a parallel-plate capacitor, the capacitance is given by

$$C = \frac{\epsilon A}{s} , \tag{6.16}$$

where ϵ is the permittivity of the insulating substance which fills the space between the conducting plates, A is the area of the plates and s is the separation between the plates. In the special case where the insulator is vacuum (i.e. empty space), $\epsilon = \epsilon_o$, the permittivity of free space.

18. The electric potential energy of a charged capacitor is given by

$$U = \frac{QV}{2} = \frac{CV^2}{2} = \frac{Q^2}{2C} . \tag{6.17}$$

19. A battery is a device which separates positively and negatively charged particles, producing a permanent electric potential difference between its terminals. The magnitude of this potential difference is called the "terminal voltage" or more often simply the voltage. If a battery is connected to a capacitor, the voltage on the capacitor will quickly become equal to the terminal voltage.

20. If two capacitors are connected in series, the electric charge stored by the first capacitor is equal to the charge stored by the second capacitor and equal to the total amount of charge stored by the combination. The voltage across the two capacitors equals the sum of the individual voltages. Thus, the effective capacitance C of two capacitors is given by

$$\frac{1}{C} = \frac{1}{C_1} + \frac{1}{C_2} . \tag{6.18}$$

21. If two capacitors are connected in parallel, the voltage across the first capacitor is equal to the voltage across the second capacitor and equal to the total voltage across the combination. The total electric charge stored by the two capacitors equals the sum of the individual charges. Thus, the effective capacitance of the combination, C, is given by

$$C = C_1 + C_2 \,. \tag{6.19}$$

22. The simplest direct current (DC) electric circuit consists of a battery connected to a "modestly conducting resistor"[a] or simply a resistor. The analysis of a DC circuit does not depend on which charge sign (positive or negative) is free to move around the circuit. It is customary to consider moving "effective positive charge," while, in fact, it is normally the negatively charged electrons that are free to move. Inside a battery in a DC circuit, an "effective chemical force" constantly "pushes" effective positive charge from the negative terminal to the positive terminal, supporting a constant value of the external terminal voltage. The battery terminals produce a constant electric field inside the modestly conducting resistor. The electric force acting on the effective positive charge inside the resistor "pushes" the charge from the positive terminal to the negative terminal through the resistor. In the steady state, the chemical force in the battery is balanced by the electric force produced by the battery terminals, and the electric force in the resistor is balanced by an "effective electric friction force." Thus, the amount of electric charge passing through any cross-sectional area per unit time remains constant.

23. The work needed for the chemical force to move a unit positive charge from the negative terminal of the battery to the positive terminal is called the EMF. EMF stands for electromotive force, but it is actually the "electromotive potential." In an ideal battery the

---

[a]The term modestly conducting resistor is used instead of conductor because the current through a good conductor can be large enough to damage both the conductor and the battery. The modestly conducting resistor could consist of, for example, a pair of conducting wires and an electric light bulb.

terminal voltage equals the EMF. In a real battery connected to a modestly conducting resistor, the terminal voltage is smaller than the EMF.

24. The electric current in a DC electric circuit is defined as the magnitude of the electric charge passing through a cross-sectional area of a circuit in a unit time. Electric current is a positive scalar, it does not have a direction. However, it is customary to use the term "direction of the current," which means the direction of motion of effective positive charge.

25. For a resistor, the electric current I is proportional to the magnitude of electric potential difference V (voltage) across the resistor ("Ohm's Law"):

$$I = \frac{V}{R}. \tag{6.20}$$

The quantity R is called the resistance of the resistor. For a resistor of fixed cross-sectional area A and length L, the resistance is proportional to L and inversely proportional to A:

$$R = \frac{\rho L}{A}. \tag{6.21}$$

The quantity ρ is called the resistivity.

26. One can calculate the energy associated with the electric current in a DC electric circuit. This energy is dissipated due to the effective electric friction in the resistor. Typically the energy of an electric current is transformed into internal (thermal) energy in the resistor. The magnitude of the rate of dissipation of electric energy (the dissipated power) is given by

$$P = IV = I^2 R = \frac{V^2}{R}. \tag{6.22}$$

Here I is the current in the circuit and V is the voltage across the resistor of resistance R. The internal (chemical) energy of the battery is continuously transformed into electric energy which is dissipated as thermal energy.

27. If two resistors are connected in series, the voltage across the two resistors equals the sum of the individual voltages. The current through the first resistor equals the current through the second resistor, and equals the total current in the circuit. Thus, the effective resistance of the two resistors is

$$R = R_1 + R_2. \tag{6.23}$$

28. If two resistors are connected in parallel, the total current in the DC circuit equals the sum of the individual currents. The total voltage equals the voltage across either resistor (both resistors have the same voltage). Thus,

$$\frac{1}{R} = \frac{1}{R_1} + \frac{1}{R_2}. \tag{6.24}$$

29. If electric currents flow through two conductors, there is a magnetic interaction between these conductors. To describe this interaction one introduces the magnetic field $\vec{\mathbf{B}}$. Every conductor carrying a current produces a magnetic field, which interacts with any other conductor carrying a current.

30. If a current I flows through a linear conducting segment $d\vec{\mathbf{L}}$, the magnetic field $d\vec{\mathbf{B}}$ produced by this segment at the point P is given by (the "Biot–Savart Law")

$$d\vec{\mathbf{B}}_P = \frac{\mu_o}{4\pi} \frac{I d\vec{\mathbf{L}} \times \vec{\mathbf{n}}_P}{r_P^2}. \tag{6.25}$$

Here μ_o is the permeability of free space, the vector $d\vec{\mathbf{L}}$ shows the direction of flow of the effective positive charge, dL is the length of the infinitesimally short conducting segment, r_P is the distance between the segment and the point P, and $\vec{\mathbf{n}}_P$ is the unit vector which points from the segment $d\vec{\mathbf{L}}$ to the point P.

31. The total magnetic field produced by a thin conducting wire at the point P can be found using

$$\vec{\mathbf{B}}_P = \int_L d\vec{\mathbf{B}}_P = \frac{\mu_o I}{4\pi} \int_L \frac{d\vec{\mathbf{L}} \times \vec{\mathbf{n}}_P}{r_P^2}. \tag{6.26}$$

where the integral is taken over the whole wire.

32. The magnetic moment $\vec{\mathbf{m}}$ of a planar current loop is defined as

$$\vec{\mathbf{m}} = \vec{\mathbf{n}} I A. \qquad (6.27)$$

Here I is the current in the loop, A is its area, and $\vec{\mathbf{n}}$ is a unit vector, perpendicular to the plane of the loop. The direction of motion of the effective positive charge in the loop is counterclockwise relative to the direction of $\vec{\mathbf{n}}$.

33. The magnetic field produced by a small planar current loop at a point P far from the loop is given by

$$\vec{\mathbf{B}}_P = \frac{\mu_o}{4\pi} \frac{3(\vec{\mathbf{m}} \cdot \vec{\mathbf{n}}_P)\vec{\mathbf{n}}_P - \vec{\mathbf{m}}}{r_P^3}. \qquad (6.28)$$

Here $\vec{\mathbf{n}}_P$ is the unit vector which points from the loop to the point P, and r_P is the distance between the loop and the point P.

34. To describe the magnetic properties of a particle one introduces the magnetic moment of the particle $\vec{\mathbf{m}}$. The magnetic field produced by the particle at a distant point P is given by the same expression as the magnetic field produced by a small current loop. For a spherical magnet this formula is valid for any point P outside the magnet.

35. The magnetic field, like an electric field, can be visualized using magnetic field lines. Unlike the electric field produced by fixed electrically charged particles, the magnetic field is a "curl" field. The magnetic field lines are closed curves. They do not have starting or ending points.

36. Ampere's Law provides an important integral expression for the magnetic field produced by an electric current

$$\oint_L \vec{\mathbf{B}} \cdot d\vec{\mathbf{L}} = \mu_o I, \qquad (6.29)$$

where the integral $\oint$ is taken over a closed line (loop) L, and current

I is the net current which passes through the surface bounded by the closed line. The direction of $d\vec{L}$ is counterclockwise relative to the direction of the net current.

37. If an electric field changes with time, it generates a magnetic field. The extended form of Ampere's Law takes this source of magnetic field into account:

$$\oint_L \vec{B} \cdot d\vec{L} = \mu_o \left(I + \epsilon_o \frac{d\Phi_e}{dt} \right) . \qquad (6.30)$$

The second term in the parentheses on the right-hand side of this expression is called the "displacement current." The vector $d\vec{L}$ is counterclockwise relative to the vector $d\vec{A}$ in the expression for the electric flux Φ_e defined by equation (6.7).

38. The magnetic force produced by a magnetic field $\vec{B}$ on a particle with electric charge q, moving with velocity $\vec{v}$ (the "Lorentz Force") is given by

$$\vec{F} = q\vec{v} \times \vec{B} . \qquad (6.31)$$

39. The magnetic force produced by a magnetic field $\vec{B}$ on a linear electric current element $Id\vec{L}$ is given by

$$d\vec{F} = Id\vec{L} \times \vec{B} . \qquad (6.32)$$

40. If a planar current loop with magnetic moment $\vec{m}$ is placed into a uniform magnetic field, the net magnetic force on the loop is zero. The torque produced by the magnetic field is given by

$$\vec{\tau} = \vec{m} \times \vec{B} . \qquad (6.33)$$

The same expression is valid for a magnetic particle of magnetic moment $\vec{m}$.

41. For both a current loop and a magnetic particle, one can introduce the magnetic potential energy associated with the direction of the magnetic moment $\vec{m}$ relative to the direction of a uniform magnetic field $\vec{B}$:

$$U = -\vec{m} \cdot \vec{B} + \text{constant.} \qquad (6.34)$$

42. The magnetic flux through a surface S is defined just like the electric flux as

$$\Phi_m = \int_S \vec{\mathbf{B}} \cdot d\vec{\mathbf{A}} \,. \tag{6.35}$$

43. If the magnetic flux changes with time it generates an electric field with a "curl." The electric field lines in this case are closed: they do not have starting or ending points.

44. "Faraday's Law of Induction" provides an integral expression relating the "curl" electric field and the magnetic flux through a surface bounded by a closed line L

$$\oint_L \vec{\mathbf{E}} \cdot d\vec{\mathbf{L}} = -\frac{d\Phi_m}{dt} \,, \tag{6.36}$$

where the integral is taken over the closed line L and the direction of $d\vec{\mathbf{L}}$ is counterclockwise relative to the direction of $d\vec{\mathbf{A}}$ in the expression for the magnetic flux Φ_m (6.35).

45. When the magnetic flux through a surface bounded by a conducting loop changes, due to a change in the magnetic field or due to motion of the conducting loop, the changing magnetic flux induces an electromotive force $\mathcal{E}$ in the loop

$$\mathcal{E} = -\frac{d\Phi_m}{dt} \,, \tag{6.37}$$

and an induced electric current $I = |\mathcal{E}|/R$, where R is the resistance of the loop. An overall positive sign for $\mathcal{E}$ means that the direction of the induced current is counterclockwise relative to the direction of $d\vec{\mathbf{A}}$ in the expression for the magnetic flux Φ_m. An overall negative sign for $\mathcal{E}$ corresponds to the opposite direction of induced current.

46. "Lenz's Law" provides an independent way to find the direction of the induced current. It states that the magnetic field of the induced current always opposes the change in the magnetic flux through the conducting loop.

47. The magnetic flux produced by a current loop (or coil) through

the surface bounded by this loop is proportional to the current in the loop

$$\Phi_m = L_m I. \qquad (6.38)$$

The coefficient L_m is called the inductance of the loop.

48. The magnetic energy stored in a current loop (or coil) is given by

$$E_m = \frac{1}{2} L_m I^2. \qquad (6.39)$$

49. Electromagnetic radiation is an electromagnetic wave consisting of propagating disturbances of the electric and magnetic fields, (i.e. electromagnetic fields) in free space or in matter. If, in a region of space the electric and magnetic fields depend only on one spatial coordinate (say x) and time, the electromagnetic wave in that region is called a plane electromagnetic wave. If the electric fields oscillates in a single plane, the wave is said to be linearly polarized, and the polarization direction is taken to be along the direction of oscillation of the electric field.

50. The simplest traveling electromagnetic wave is a plane, linear polarized harmonic wave, in which the electric field oscillates along one axis (say, the y-axis), the magnetic field oscillates along another axis (say z-axis), and the wave propagates along the third axis (x-axis). We consider only this simplest case below.

51. The electric and magnetic fields in the simplest electromagnetic wave are given by

$$E_y = E_o \cos(kx - \omega t + \phi), \quad B_z = B_o \cos(kx - \omega t + \phi), \quad (6.40)$$

if the wave travels in the positive x-direction and

$$E_y = E_o \cos(kx + \omega t + \phi), \quad B_z = -B_o \cos(kx + \omega t + \phi), \quad (6.41)$$

if the wave travels in the negative x-direction. Here k is the wave number, ω is the angular frequency, and ϕ is the phase constant of the wave.

52. The speed of an electromagnetic wave traveling in a vacuum is commonly designated as c:

$$c = \frac{\omega}{k} = \frac{\lambda}{T} = \lambda f, \qquad (6.42)$$

where λ is the wavelength, T is the period and f is the frequency of the wave. In free space, the value of c is related to the permittivity ϵ_o and the permeability μ_o:

$$c = \left(\epsilon_o \mu_o\right)^{-\frac{1}{2}}. \qquad (6.43)$$

The ratio of the amplitudes of the electric and magnetic fields in an electromagnetic wave is equal to the speed c:

$$\frac{E_o}{B_o} = c. \qquad (6.44)$$

53. The energy density associated with the electric field $\vec{\mathbf{E}}$ and the magnetic field $\vec{\mathbf{B}}$ are given by

$$u_E = \frac{\epsilon_o E^2}{2}, \quad u_B = \frac{B^2}{2\mu_o}. \qquad (6.45)$$

54. The average energy density (over a period) in a region of electromagnetic fields is

$$u_{av} = \frac{E_o B_o}{2\mu_o c}. \qquad (6.46)$$

55. The power per unit area carried by an electromagnetic wave through a perpendicular surface is given by the magnitude of the "Poynting vector" $\vec{\mathbf{S}}$

$$\vec{\mathbf{S}} = \frac{\vec{\mathbf{E}} \times \vec{\mathbf{B}}}{\mu_o}. \qquad (6.47)$$

The direction of the Poynting vector is the direction of propagation of the wave.

56. The intensity of an electromagnetic wave, (i.e. electromagnetic

radiation) is the average (over a period) of the Poynting vector magnitude

$$I_r = \frac{E_o B_o}{2\mu_o} \,.$$

(6.48)

57. The radiation pressure P_r is the average force per unit area (magnitude) exerted by an electromagnetic wave incident on an object. If the wave is incident perpendicular to a totally absorbing surface, the radiation pressure is

$$P_r = \frac{I_r}{c} \,.$$

(6.49)

58. If an electromagnetic wave is incident on an object whose size is much greater than the wavelength, one can use the "ray approximation." This means the waves propagate along straight lines (called rays), which can change their direction only at the boundary between two substances. The ray approximation is often used for the visible part of the electromagnetic spectrum ("ordinary visible light").

59. If a parallel beam of light falls on a smooth flat surface, the angle of incidence is equal to the angle of reflection, where the angles are measured from the normal to the surface (the "Law of Reflection," see Fig. 8).

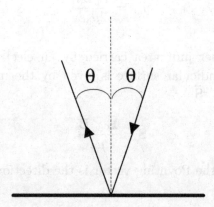

Fig. 8.   Reflection of light in the ray approximation.

60. If light passes through the boundary between two substances the angle of refraction θ_2 is connected to the angle of incidence θ_1 by "Snell's law" (see Fig. 9):

$$n_1 \sin \theta_1 = n_2 \sin \theta_2 \,, \tag{6.50}$$

where n_1 is the refractive index of the first substance and n_2 is the refractive index of the second substance. The refractive index of a substance is given by $n_i = c/v_i$ where c and v_i are the speeds of light in free space and in the substance of interest respectively.

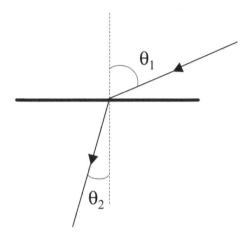

Fig. 9.   Refraction of light in the ray approximation.

61. If a linearly polarized electromagnetic wave of intensity I_r passes through a polarizing sheet, the intensity of the transmitted light I_r' is given by "Malus's Law":

$$I_r' = I_r \cos^2 \theta \,, \tag{6.51}$$

where θ is the angle between the electric field of the incident electromagnetic wave and the transmission axis of the polarizing sheet. For the transmitted light, the electric field oscillates along the transmission axis of the polarizing sheet, and the magnetic field is perpendicular to the electric field.

6.2.   Problems

1. Estimate the magnitude of the electric field produced by an up quark at a distance of 2 fm from the quark. Use the data in Table 2 in Chapter 2.

2. Estimate the electric potential produced by an up quark at a distance of 2 fm from the quark, assuming zero electric potential at infinity.

3. Estimate the magnitude of the electric force between an up quark and a down quark if the distance between them is 1 fm.

4. A tau lepton was created in a uniform electric field of magnitude 1 MV/m. Estimate the mean distance traveled by the tau lepton during its lifetime if its initial velocity is 10 km/s pointing in the direction of the electric field. Assume the mean lifetime of the tau lepton to be 290 fs. Use the data in Table 1 in Chapter 2.

5. Solve problem 4 if the initial velocity points in the direction opposite to the direction of the electric field.

6. A tau lepton was created between the plates of a parallel-plate capacitor of area 1 cm^2 and separation 1 mm. Estimate the magnitude of the tau lepton's acceleration if the charge on the capacitor is 1 μC. What is the direction of the acceleration relative to the plates?

7. A muon was created in a region with a uniform magnetic field of magnitude 10 T. If the initial velocity of the muon is 100 km/s perpendicular to the direction of the magnetic field, find the magnitude of its acceleration. Use the data in Table 1 in Chapter 2.

8. In problem 7, find the mean distance traveled by the muon during its lifetime. Take the mean lifetime of the muon to be 2.2 μs.

9. Estimate the electric potential energy of the three quarks in a proton. (Take the potential energy to be zero when the distance be-

tween the quarks is infinite). Assume that the quarks are stationary and located at the vertices of an equilateral triangle of side 1.2 fm and use Tables 2 and 3 in Chapter 2 for the other data needed.

10. Solve problem 9 for a neutron.

11. Find the magnitude of the acceleration of a tau lepton located at a distance of 1 nm from a proton, assuming that the only force acting on the tau lepton is the electric force produced by the proton.

12. Solve problem 11 for a muon.

13. Estimate the magnitude of the electric field produced by a ^3_2He nucleus at a distance of 1 nm from its center.

14. In problem 13 estimate the electric potential produced by the nucleus. Take zero potential to be at an infinite distance from the nucleus.

15. Estimate the maximum possible electric force (magnitude) between two stable nuclei at a distance of 1 nm between their centers.

16. Estimate the magnetic potential energy difference for two states of a ^3_2He nucleus if: (1) the magnetic moment of the nucleus points in the direction of a magnetic field of magnitude 1 T, and (2) the magnetic moment of the nucleus points in the direction opposite to the direction of the magnetic field. (The magnetic moment of ^3_2He is 2.1 μ_n.)

17. In problem 16 find the maximum possible torque (magnitude) produced by the magnetic field on the nucleus.

18. A $^{209}_{83}Bi$ nucleus is placed at the origin. Its magnetic moment ($m = 4\mu_B$) points in the positive z-direction. Find the magnitude and direction of the magnetic field produced by the nucleus at the point $x = y = 0, z = 1$ nm.

19. In problem 18 find the magnitude and direction of the magnetic field at points
   a) $x = y = 0, z = -1$ nm;
   b) $y = z = 0, x = 1$ nm.

20. Deuterium and tritium can fuse producing helium. Estimate the force of electric repulsion (magnitude) between deuterium ^2_1H and tritium ^3_1H nuclei with a distance of 10 fm between their centers.

21. In problem 20, how much energy is needed to reach a distance of 10 fm between the two nuclei if the initial distance was very much greater than 10 fm?

22. In field emission transmission electron microscopy (FETEM) electrons are drawn out of a cathode by a high voltage. In a modern device, the electric potential difference used can be as high as 1 MV. Assuming that the initial speed of the electrons is small compared to the final speed, estimate the final kinetic energy of the electrons (in joules) accelerated by the 1 MV voltage.

23. To compress nuclei and produce a new form of nuclear matter, scientists smash a high energy beam of lead ions into a fixed target of lead or gold atoms. During a collision, the maximum nuclear matter density is estimated to be 20 times the normal nuclear density. Assume that two lead nuclei collide, producing a "fireball" of 20 times normal nuclear density. Estimate the radius of the fireball in terms of the radius of a lead nucleus, and the magnitude of the electric field at the surface of the fireball in terms of the magnitude of the electric field at the surface of the lead nucleus. Assume that the mass of the fireball is approximately double the mass of the lead nucleus.

24. In one of the versions of magnetic resonance force microscopy (MRFM) a cantilever with a small magnet on its tip oscillates near a sample causing cyclic change in the direction of the magnetic moment of an atom in the sample. In turn, the atomic magnetic moment reversals cause a change in the cantilever oscillation frequency, which can be detected. Assume that both the magnetic moment of the mag-

net and the atomic magnetic moment in the sample point along the line connecting them. Estimate the magnetic field (magnitude) produced by the magnet at the atom and magnetic field produced by the atom at the magnet, if they are separated by the distance 1 μm. Take the magnetic moment of the magnet to be 1.5×10^{-12} Am^2, and the magnetic moment of the atom to be 9.3×10^{-24} Am^2.

25. In problem 24 assume that the cantilever oscillates (vibrates) with an amplitude of 10 nm. Estimate the range of the change in magnetic field at the atom.

26. The intrinsic magnetic moment of the muon m_μ has been measured recently with an accuracy of about 1.3 parts per million. The precise value of m_μ is extremely important for verification of the theory of fundamental elementary particles. Assume that the measured magnetic moment of the muon is greater than the theoretically predicted value by 2 parts per million. Let a muon be placed in a high magnetic field with $B = 20$ T.

a) Estimate the difference between the theoretical and experimental values of intrinsic magnetic moment (take the theoretical value to be 9.3×10^{-24} J/T).

b) Estimate the corresponding shift in the difference between the maximum and minimum magnetic energies of the muon in the 20 T magnetic field.

27. A "Halbach Array" of permanent magnets is shown in Fig. 10.

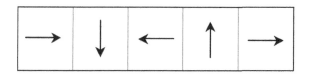

Fig. 10.   A Halbach array. The arrows show the directions of the individual magnetic moments.

The arrows in the figure show the direction of the magnetic moment (from south pole to north pole). On which side of the Halbach array (top or bottom) do the magnetic fields from neighboring magnets reinforce each other? On which side do they tend to cancel each other?

28. In a scanning tunneling microscope, a tungsten tip is brought very close to the conducting surface under investigation. A small tunneling current of about 2 pA flows from the tip to the surface (electrons "jump" from the surface to the tip). Consider a system of coordinates with the origin on the conducting surface and z-axis directed along the tip (see Fig. 11).

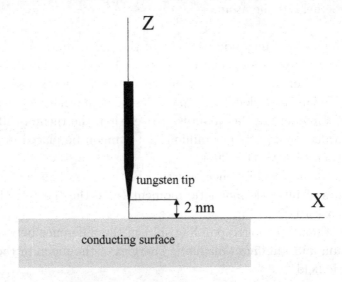

Fig. 11.   A tungsten tip near a conducting surface.

Find the magnetic field (in unit vector notation) produced at the point P ($x_P = 36$ nm, $y_P = 64$ nm, $z_P = 0$), by the current segment flowing between the tip and the surface. Take the separation between the tip and the surface to be 2 nm.

29. In problem 28 consider a paramagnetic atom placed at the point P. The magnetic moment of the atom has a magnitude of one Bohr magneton (μ_B) and points in the positive x-direction.

a) Find the torque (in unit vector notation) produced by the magnetic field of the tunneling current on the atom.

b) Find the change of the atomic magnetic energy due to the interaction with the tunneling current, if the direction of the atomic magnetic moment reverses.

30. In problem 28 consider the probe and the sample as a capacitor with a capacitance of 10^{-16} F. If the voltage between the tip and sample is 1 V, find the induced tip charge.

31. It is estimated that conducting nanotubes can withstand electric current densities of up to 10^{13} A/m^2, 1000 times greater than copper. Consider a conducting wire made of carbon nanotubes. Let the current density in the wire $j = 1$ TA/m^2 (T here stands for tera $= 10^{12}$). If the cross-sectional area of the wire is 1 mm^2, estimate the charge which passes through the cross-sectional area in 10 seconds.

32. Consider a small 1 cm segment of the wire discussed in problem 31, located at the origin of coordinates. If the current flows in the positive z-direction, find (in unit vector notation):
    a) the magnetic field produced by the segment at the point P (10 cm, 15 cm, 20 cm);
    b) the magnetic force produced by the segment on an electron at the point P, moving at 10^5 m/s in the positive z-direction.

33. Consider a circular current loop of radius 1 cm made of the wire discussed in problem 31.
    a) What is the magnetic moment (magnitude) of the current loop?
    b) If the current loop is placed at the origin, and the current flows in the x–y plane in a counterclockwise direction relative to the z-axis, find the magnetic field at the point P (10 cm, 15 cm, 20 cm) in unit vector notation.
    c) Find the torque on an atom (in unit vector notation) at the point P if the atom has a magnetic moment of two Bohr magnetons, which points in the positive z-direction.

34. Consider a square current loop of side 1 cm made of the material described in problem 31, which is placed in a uniform magnetic field of 1 T. The plane of the loop is perpendicular to the magnetic field.

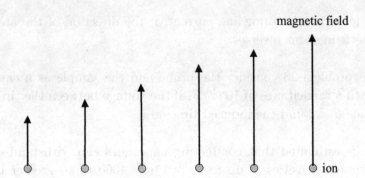

Fig. 12.   A nonuniform magnetic field on a chain of ions.

a) What is the change of the magnetic energy of the loop (absolute value) if the direction of the current reverses?

b) What is the torque (magnitude) on the loop produced by the magnetic field?

35. One of the proposals for quantum computation suggests using a one dimensional chain of tellurium ions on a silicon surface. To allow manipulations with single ions, the sample is placed in a nonuniform magnetic field. Assume that the magnetic field (magnitude) at the left end of the chain is 5 T. The magnetic field for each ion is 5×10^{-4} T greater than the magnetic field at its neighbor to the left. (See Fig. 12, where gray dots indicate tellurium ions and the arrows show the magnetic field on the ions.)

Let the chain contain 100 ions. Consider two states of an ion: (1) the magnetic moment of the ion points in the direction of the magnetic field, and (2) the magnetic moment points in the opposite direction. Find the difference in magnetic energy for these two states:

a) for the ion at the left end;

b) for its neighbor;

c) for the ion at the right end.

The magnetic moment of the tellurium ion equals one Bohr magneton (μ_B).

36. In problem 35 consider the same chain of tellurium ions but in a uniform magnetic field. To "select" an ion, a small spherical magnet

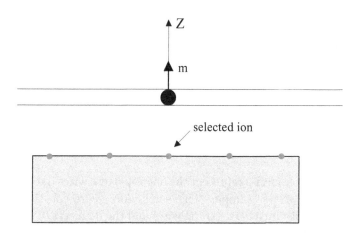

Fig. 13. A movable magnet "selects" an ion on the surface of a sample.

moves along the chain and "stops" exactly above the selected ion (see Fig. 13).

The quantity $\mu_0 m/V$ is called the magnetic induction of the magnet, where m is the magnetic moment of the magnet, and V is its volume. Assume that the magnetic induction is 2.2 T, the radius of the magnet is 5 nm, and the distance between the center of the magnet and the selected ion is 15 nm. Find:

a) the magnetic field produced by the magnet on the selected ion;

b) the maximum possible torque (magnitude) produced by the magnet on the selected ion;

c) The maximum change in the magnetic energy of the selected ion connected to the direction of its magnetic moment.

37. One of the proposals for quantum computation suggests using a chain of complicated "molecules," on a silicon surface. Such molecules can be precisely positioned on a silicon surface using scanning tunneling microscopy. C_{60} molecules can be used as small (≈ 1 nm) traps for paramagnetic nitrogen atoms, designated as $N@C_{60}$. To create a nonuniform magnetic field at the nitrogen atoms, it was suggested that so-called "micro-patterned wires" be used. Assume that two wires are placed on the silicon surface as shown in Fig. 14.

Assume that an electric current of 1 A flows in both wires in the positive y-direction (into the page), the distance between the wires

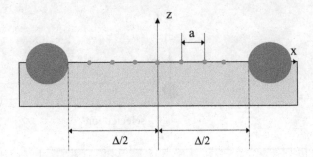

Fig. 14.   The large circles represent the micro-pattern wires (cross-sections) on the surface of the sample. The small dots represent the chain of molecules. The distance between the wires and the diameter of the wires in the Fig. are not to scale.

$\Delta = 1$ μm, and the diameter of a wire $d = 1$ μm. Find the magnetic field (in unit vector notation) produced by the wires:

   a) at the origin (i.e. the midpoint between the wires);

   b) at the nitrogen atom at $x = a$ (take the distance between the neighboring nitrogen atoms $a = 1.14$ nm);

   c) at $x = -a$.

38. In problem 37 consider three neighboring nitrogen atoms at points $x = 0, x = a$, and $x = 2a$. Find the difference in the magnetic energy for these atoms assuming that all three magnetic moments point in the positive z-direction. Take the magnetic moment of the nitrogen atom to be $3\mu_B$.

39. The luminosity (the radiated light power) of the Sun is about 4×10^{26} W. Find the energy emitted by the Sun in 1 year.

40. In problem 39 find the intensity of the Sun's electromagnetic radiation at a distance of 0.1 AU from the center of the Sun (1 AU $\approx 1.5 \times 10^{11}$ m).

41. The orbit of the asteroid Aten has a semimajor axis of 0.97 AU. Estimate the maximum radiation pressure on the surface of the asteroid when the distance between the asteroid and the center of the Sun is 0.9 AU. (The radiation pressure is maximum for the surface,

which is perpendicular to the incident light.) The light power radiated by the Sun is 4×10^{26} W. Assume that Aten's surface absorbs all the light falling on it.

42. Space-based x-ray observations hint at the existence of magnetars (neutron stars which produce magnetic fields of up to 10^{11} T). Assume that an electron is traveling at 100 km/s near the surface of a magnetar. What is the maximum magnetic force (magnitude) that can be exerted on the electron, if the magnetic field is 10^{11} T?

43. In problem 42 what maximum torque (magnitude) can be exerted on the intrinsic magnetic moment of the electron, which is equal to one Bohr magneton?

44. It was shown that pulses of hard x-ray radiation can be trapped in a crystal resonator: a pair of thin slabs made from silicon and placed about 15 cm apart. Let a 0.1 ns x-ray pulse of frequency about 4×10^{18} Hz travel back and forth between the slabs. Find:
    a) the wavelength of the x-ray;
    b) the period of the x-ray;
    c) the length of the pulse;
    d) the number of wavelengths in the pulse.

45. In problem 44, assume that after every reflection the energy of the pulse decreases by 30%.
    (a) How many reflections does the pulse experience before it loses more than 90% of its initial energy?
    (b) Estimate the distance traveled by the pulse before it loses more than 90% of its energy.
    (c) How long does it take the pulse to cover the distance found in part (b)?

46. Optical standing waves between two mirrors can trap neutral atoms at the antinodes of the wave. If the frequency of the light is 3.5×10^{14} Hz what is the distance between two neighboring traps?

47. In experiments on "stopping light," scientists irradiated a

Rubidium–Helium gas mixture with a "control beam," which changes the properties of the gas. In the presence of the control beam, they sent a "signal light pulse" into the gas from another source of light whose duration is τ. Inside the gas the speed of the signal pulse drops to approximately 900 m/s. When the signal was "compressed" inside the gas, the control beam was turned off. The light inside the gas (both "control" and "signal") disappeared. When the scientists irradiated the gas with the control beam again, the gas emitted a signal light pulse of the same duration τ. In other words the signal pulse was essentially "stopped and then released" by the gas.

a) Taking $\tau = 20$ μs, find the length of the signal pulse before entering the gas.

b) Find the same length inside the gas.

48. Our Milky Way Galaxy has two small satellite galaxies which are called the "Magellanic Clouds." The smaller of these (the "Small Magellanic Cloud") lies approximately 200,000 ly away from the Sun. The space-based Hubble Space Telescope routinely produces sharp images of stars in the Small Magellanic Cloud. Assuming that the light detected by the Hubble telescope has a wavelength 600 nm, estimate the number of wavelengths between a star in the Small Magellanic Cloud and the Hubble telescope.

49. Galaxy clusters are the largest "structural elements" in the universe. Hot gas typically fills the space within a cluster. The hot gas radiates x-rays, which can be detected by "x-ray telescopes." In 2002 the distant cluster of galaxies "3C294" was discovered by the space-based Chandra X-Ray Observatory. Its size is estimated to be 6×10^5 ly. If the angular size of the cluster is about 6×10^{-5} rad estimate how long ago the x-rays detected by Chandra were emitted by the cluster.

50. A composite material, made of fiberglass and copper, has a "negative refractive index" for microwave radiation of frequency about 5 GHz. The negative refractive index in Snell's Law means that the angle of refraction is negative, i.e. the refractive wave propagates on the same side of the normal as the incident wave.

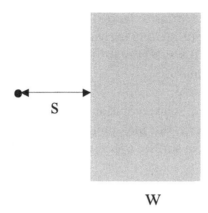

Fig. 15. A "point source" of microwaves is placed at a distance S from a slab of width W ($W > S$).

   a) What is the wavelength of microwaves with frequency 5 GHz?

   b) Let a point source of microwaves be placed at a distance S from a slab of a composite with refractive index $n = -1$ and width W. (See Fig. 15.)

Find the position of the image of the point source inside the slab (graphically). What is the distance between the object and the image?

   c) Find the position of the final image outside the slab. What is the distance between the object and the image?

51. The Solar Monitoring Observatory (SOLAR) on the International Space Station (ISS) studies the Sun's radiation in the region of wavelengths between 17 nm and 100 μm. Find the corresponding range of frequencies of solar radiation.

52. The Monitor of All-sky X-ray Image (MAXI) on the ISS detects the x-ray radiation in the wavelength region from 42 pm to 2.5 nm. Find the minimum and maximum oscillation periods of the x-ray radiation detected by the MAXI.

53. The Alpha Magnetic Spectrometer (AMS-02) on the ISS produces a maximum magnetic field of 750 mT. Find the maximum magnetic force exerted by AMS-02 on a proton of energy 1 eV.

54. When the rover Curiosity landed on the surface of Mars in 2012, the information about landing arrived at the Earth 14 minutes after it was sent by the rover. Estimate the distance between the Earth and Mars at that time.

55. The intensity of solar radiation at the Earth's orbit is about 1360 W/m^2. Estimate the maximum force the solar radiation can exert on the reflective surface of a solar sail at the Earth's orbit if its area is 1 km^2.

56. In Problem 55 estimate the maximum force on the same sail at Uranus' orbit (about 19 AU).

57. The amplitude of an alternating electric field in a linear accelerator is about 30 MV/m.
   a) Estimate the corresponding maximum electric energy density.
   b) Estimate the maximum electric force on an electron in that field.

58. The maximum intensity of an x-ray laser's radiation is about 1022 W/m^2.
   a) Estimate the corresponding average energy density of the radiation.
   b) If the laser radiation is absorbed by an object what is the radiation pressure on the object?
   c) Estimate the amplitude of the electric and magnetic fields in the x-ray beam.

59. In the Large Hadron Collider (LHC), protons travel inside the 27-km long circular tunnel with a speed very close to the speed of light. The force required to keep the protons in this circular orbit is estimated to be about 4×10^{-10} N. Find the magnetic field required to keep the protons in the circular LHC orbit of this radius. (The magnetic field is perpendicular to the velocity of the protons.)

60. In order to study the electric currents in photosynthesis, scientists insert nanoelectrodes into a living plant cells less than 10 microme-

ters in diameter. Assume that under the action of a light pulse, the current in the cell is 1.5 pA. If the bias voltage is 200 mV and the duration of the light pulse is 10 s, find the electric energy dissipated in the cell during the light pulse.

Chapter 7

Hints to Solution

7.1. Mechanics

1. The gravitational acceleration near the ground is $g = GM_e/R_e^2$. At the height h the gravitational acceleration is $g' = GM_e/(R_e + h)^2$. The relative change of gravitational acceleration in percent is

$$\frac{g - g'}{g} \times 100 \approx \frac{200h}{R_e} .$$

2. The gravitational force at a height h is given by

$$F_g = GM_e m/(R_e + h)^2 = mg\left(\frac{R_e}{R_e + h}\right)^2 .$$

where m is the mass of the bacteria.

3b. $s = v\Delta t$ where v is the speed of the probe and $\Delta t = 1$ year.

4b. $F = F_g = GM_e m/r_{es}^2$, where r_{es} is the distance between the center of the Earth and the spacecraft and m is mass of the spacecraft.

4c. $GM_e m/r_{es}^2 = GM_o m/r_{ss}^2$, where r_{ss} is the distance between the center of the Sun and spacecraft, and r_{es} is the distance between the center of the Earth and the spacecraft.

5a. $s = v\Delta t$.

5b. $p = mv$, $K = mv^2/2$.

6a. $F_g = GM_e m/(r_e^2 + h)^2 = mg\left(\dfrac{R_e}{R_e + h}\right)^2.$

6b. $a = F/m \approx F_g/m = g\left(\dfrac{R_e}{R_e + h}\right)^2.$

7. $v = s/\Delta t.$

8. The gravitational force (magnitude) exerted by the Sun F_{gs} is $F_{gs} = GM_o m/r_{ss}^2$, where r_{ss} is the distance between the centers of the Sun and the spacecraft. The gravitational force exerted by the Earth $F_{ge} = GM_e m/r_{es}^2$ where r_{es} is the distance between the center of the Earth and the spacecraft. The ratio is

$$\frac{F_{gs}}{F_{ge}} = \frac{M_o}{M_e} \cdot \left(\frac{r_{es}}{r_{ss}}\right)^2.$$

9a. Consider a system of coordinates connected to the spacecraft. For a small time interval (the time of the collision) it can be treated as an inertial system. The magnitude of the impulse I_a delivered by the spacecraft to the atom is $I_a = mv$, where v is the speed of the atom. It follows from the Action–Reaction Principle that the magnitude of the impulse delivered by the atom to the spacecraft I_s is equal to I_a. To estimate the mass of the gold atom, use the Periodic Table of Elements. The atomic mass of Au atom is approximately 197. Thus 1 mol of gold, which contains Avagadro's number of atoms, has a mass of 197 g.

9b. Yes, approximately. During the small time of collision we can neglect the external gravitational forces, so the total momentum is conserved.

9c. $K = mv^2/2$. No. Kinetic energy is not conserved in inelastic collisions (in our problem we have a perfectly inelastic collision: the atom sticks to the collector).

10a. $\vec{L} = \vec{r} \times \vec{p} = m\vec{r} \times \vec{v} = 0.$

10b. $\vec{L} = m\vec{r} \times \vec{v}, \quad L = mrv \sin \phi = mR_e v.$

11. $F_g = GM_o m/r^2$, where m is the mass of the spacecraft and r is the distance between the Sun and Saturn.

12. $g_T = GM_T/R_T^2$, where M_T and R_T are the mass and radius of Titan respectively.

13. $F_g = GM_J m/r^2$, where r is the distance between the spacecraft and Jupiter's center, M_J is the mass of Jupiter, and m is the mass of the spacecraft.

14a. $p = mv, \quad K = mv^2/2.$
To find the mass of an ammonia molecule NH_3 find its molar mass, i.e. the mass of Avogadro's number of molecules.

14b. $L \approx mR_J v.$

15. $M = \rho V$, where ρ is the density of ice, $V = 4\pi R^3/3$ is the volume of the comet's nucleus.

16. $\rho = M/V$, where M is the mass and V is the volume of the coma.

17a. $S/\Delta t = 4d_a/T$, where d_a is the Sun–comet distance at aphelion, and T is the period of the comet.

17b. $F_a = GM_c M_o/d_a^2, \quad F_p = GM_c M_o/d_p^2, \quad F_a/F_p = (d_p/d_a)^2$, where F_a and F_p are the gravitational forces at aphelion and perihelion, and d_a and d_p are the Sun–comet distances at aphelion and perihelion.

18. $M = \rho V$, where M and V are the mass and volume of a fragment.

$$K = Mv^2/2, \quad p = Mv, \quad I = p.$$

19. $F_1 = GmM_J/(r + d)^2$, $F_2 = GmM_J/r^2$, where $r = 20$ mm, $d = 5$ km, and M_J is the mass of Jupiter.

$$F_2 - F_1 \approx 2GmM_J d/r^3 .$$

20a. $\rho = m/V$, $V = 4\pi r^3/3$, $r = 5$ mm.

20b. $a = F_g/m = GM_e/(R_e+h)^2 = gR_e^2/(R_e+h)^2$, where $h = 90$ km.

20c. $p = mv$, $K = mv^2/2$.

22. $E = P\Delta t$, $P = 40$ kW/14.

23a. $v = \omega_e(R_e + h)$, where ω_e is the angular speed of the Earth, $\omega_e = 2\pi/T_e$, and $T_e = 1$ day.

23b. $a = v^2/(R_e + h)$.

23c. $F_g = GM_em/(R_e+h)^2$, $F = ma$, $F = F_g - F_a$, $F_a = F_g - F$, where F_a is the magnitude of the force exerted by the air.

23d. $L = rp$, $L/p = r = (R_e + h)$.

24. $K = mv^2/2$, $p = mv$, $K/p = v/2$.

25. The proton's density is $\rho = m/V$, where m_p is the mass of the proton, $V = 4\pi r_p^3/3$, and r_p is the radius of the proton.

26. $K = (\rho V)v^2/2$, where ρ is the density of air, V its volume ($V = 1$ m^3), and v is the speed of the moving air.

27. $p = (\rho V)v$.

28. $K = mv^2/2$. To find the mass of a nitrogen molecule, N_2, use the Periodic Table of Elements. 1 mol (28 g) of molecular nitrogen contains Avogadro's number of molecules.

29a. $p = mv, \quad K = mv^2/2.$

29b. $I = p.$

30. We consider the motion of Alpha in the system of coordinates connected to the center of the Earth.

$$\frac{GM_e m}{r^2} = \frac{mv^2}{r}, \quad r = R_e + h.$$

From this equation
$$v = \sqrt{GM_e/r}, \quad \omega = v/r, \quad T = 2\pi/\omega = 2\pi r/v, \quad a = GM_e/r^2.$$

31. $GM_e m/r^2 = mv^2/r, \quad v = \sqrt{GM_e/r}.$ If r increases 9 times, v decreases $\sqrt{9} = 3$ times. $\omega = v/r.$ If v decreases 3 times and r increases 9 times, then ω decreases 27 times. The acceleration $a = v^2/r$, decreases 81 times. (You can also use $a = GM_e/r^2$.) The period $T = 2\pi/\omega$, increases 27 times.

32. $p = mv$, decreases $\sqrt{9} = 3$ times (see problem 31). $L = rp$, increases 3 times. $K = mv^2/2$ decreases 9 times.

33. $\rho = m/V, \quad m = w/g$, where w is the weight of Destiny.

34a. $F_g = GM_e m/r^2, \quad r = R_e + h.$

34b. $p = mv$ (see problem 30).

$$L = rp, \quad K = mv^2/2 = p^2/2m.$$

35. $F_1 = GM_e m/r^2, \quad F_2 = GM_o m/r_{es}^2, \quad F_1/F_2 = M_e r_{es}^2/M_o r^2,$ where $r = R_e + h$, and r_{es} is the mean distance between the Earth and the Sun.

36. $GM_e m/r^2 = mv^2/r$, where r is the radius of the geosynchronous orbit. Substitute $r = v/\omega_e$, where $\omega_e = 2\pi/T_e$ is the angular speed of the Earth, $T_e = 1$ day. Thus, $v = (GM_e \omega_e)^{\frac{1}{3}}, \quad p = mv, \quad L = rp, \quad K = mv^2/2.$

37. $T^2/A^3 = 1(year)^2/(AU)^3$ according to Kepler's Third Law.

38. $V = 4\pi r^3/3$, where r is 10^{10} ly.

39b. $v = \omega r$, $r = 50$ km.

39c. $L = I\omega$, $I = 2Mr^2/5$, $M = \rho V$.

39d. $K = I\omega^2/2$.

41a. $M_c = \rho V$, $V = \pi \cdot 0.4^2 \cdot 1.8$ km^3.

41b. $F_g = GM_eM_c/r^2$.

41c. $\omega = 2\pi/T$.

42. $\dfrac{F_I}{F_D} = \dfrac{M_I}{M_D} = \dfrac{V_I}{V_D}$, where the subscripts "$I$" and "$D$" stand for Ida and Dactyl.

43. $\rho = m_g/V_g$, where m_g and V_g are the mass and the volume of a typical dust grain.

44a. $d = 100$ m.

44b. $V_g/10^6$.

44c. $m_g/10^6$.

44d. $F_g = Gm_g^2/d^2$.

45. $\omega = 2\pi f$, $T = 1/f$, $m = k_s/\omega^2$.

46a. $x = A\cos(\omega t + \phi)$, $v_x = -A\omega\sin(\omega t + \phi)$. At $t = 0$, $x_i = A\cos\phi$, $v_{xi} = -A\omega\sin\phi$. At the end point $v_{xi} = 0$, so $\phi = 0$, or $\phi = \pi$. We choose $\phi = 0$ as $x_i > 0$. Thus, $A = x_i = 40$ nm, and the x-coordinate of the cantilever tip is given by $x = A\cos\omega t$ where A and ω are known. Substituting $t = 1$ ms find the corresponding position of the cantilever tip. For the velocity $v_x = -A\omega\sin\omega t$. For the acceleration $a_x = -A\omega^2\cos\omega t$ or one can use $a_x = -\omega^2 x$.

46b. $t = T/4$, $t = 3T/4$.

46c. Consider the equation $x = A\cos\omega t = 20$ nm. It follows that $\cos\omega t = 1/2$. Solutions are $\omega t = \pi/3+2\pi k$ and $\omega t = 5\pi/3+2\pi k$, $k = 0, 1, 2, \dots$. The first passage through the point $x = 20$ nm corresponds to the phase $\omega t = \pi/3$, the second passage corresponds to the phase $5\pi/3$. Dividing by ω, find corresponding time.

47a. $v_{\max} = A\omega$.

47b. Use conservation of mechanical energy:

$$\frac{1}{2}k_s x^2 + \frac{1}{2}mv^2 = \frac{1}{2}k_s A^2.$$

Dividing by the mass m one gets

$$(\omega x)^2 + v^2 = (\omega A)^2, \quad v^2 = \omega^2(A^2 - x^2).$$

47c. $a_{\max} = \omega^2 A$, $F_{\max} = k_s A$.

47d. $a = \omega^2|x|$, $F = k_s|x|$.

48. In the equilibrium position the effective spring force balances the magnetic force: $k_s|\Delta x| = F_m$.

49. $v_f = a t_f$, $t_f = v_f/a$, where v_f and t_f are the final speed and time.

50. $v_f^2 = 2as$, $s = v_f^2/2a$.

51. $F_m = mg$, where F_m is the magnitude of the magnetic force and m is the mass of the train car.

52. $P = F_a v$, $F_a = P/v$, where P is power, and F_a is the force of air resistance.

53. $K = mv^2/2$, $p = mv$.

54. $\frac{dp}{dt} = ma$, dp/dt is the magnitude of the net force acting on the train car.

55a. $v_f^2 = 2as$, $a = v_f^2/2s$.

55b. Let the x-axis point down the incline.

$$F_x = mg\sin\theta - F_e, \quad F_x = ma_x = -ma,$$

$$mg\sin\theta - F_e = -ma, \quad F_e = m(g\sin\theta + a),$$

where F_e is the external force.

56. $s = ((v_i + v_f)/2)t_f = v_f t_f/2$, $\quad t_f = 2s/v_f$.

57a. $v_f = at_f$, where $t_f = 1$ s.

57b. $v_f^2 = 2as$, where $s = 100$ m.

58a. Use the Work–Kinetic Energy Theorem $W_e + W_g + W_r = mv_f^2/2$, where W_e is the work done by the external force, which is equal to energy needed, W_g is the work done by the gravitational force, $W_g = U_i - U_f = -mgs\sin\theta$, and W_r is the work done by the resistive forces (energy dissipation).

58b. $P_{av} = W_e/t_f$, where t_f is the total time of acceleration, which was found in problem 56.

59. Let us start with 59b. At the end of the first second, $v_f = at_f$, where $t_f = 1s$. After the driving force is shut down $v_f' = v_i' - a'(t_f' - t_f)$, where $v_f' = 0$, $v_i' = v_f$, $a' = g\sin\theta$. Thus, $t_f' = t_f + v_f/g\sin\theta$.

59a. The distance covered during the first second, $s = at_f$, where

$t_f = 1s$. The total distance traveled $s' = s + ((v'_i + v'_f)/2)(t'_f - t_f)$, where $v'_i = v_f$, $v'_f = 0$.

60a. First consider the motion up the incline. During the first 10 m of acceleration:

$$F_x = -F_e + mg\sin\theta + \mu F_n\,,$$

where $F_n = mg\cos\theta$, F_e is the external driving force (magnitude), and the x-axis is assumed to point down the incline. Next,

$$F_x = ma_x = -ma\,.$$

Thus,

$$F_e - mg\sin\theta - \mu mg\cos\theta = ma\,,$$

and we can compute the acceleration a for the first 10 m of travel. The speed of the launcher after the first 10 m, can be found from $v_f^2 = 2as$, where $s = 10$ m.
   Now let us consider the motion after the driving force is shut down ($t = 0$) until the moment when the launcher stops. From equation $F'_x = mg\sin\theta + \mu mg\cos\theta = ma'$ we can find the magnitude of the acceleration a', which points down the incline. Next

$$v'_f = v'_{x_i} + a'_x t_f, \quad 0 = -v_f + a't'_f, \quad t'_f = v_f/a'\,,$$

$$s' = \frac{v'_i + v'_f}{2}t'_f = \frac{v'_i t'_f}{2} = \frac{v_f t'_f}{2}\,,$$

where t'_f is the time when the launcher stops and s' is the distance covered during that time. For the motion down the incline we have

$$F_x = mg\sin\theta - \mu mg\cos\theta = ma'', \quad (v''_f)^2 = 2a''s''\,,$$

where $s'' = s' + 10m$. Finally, $s'' = (v''_f/2)(t''_f - t'_f)$. Thus, $t''_f = t'_f + 2s''/v''_f$.

60b. The distance is $s' + s'' = 2s' + 10$ m.

60c. 10 m

61a. $\Delta t = s/v$, where $s = 4000$ km $= 4 \times 10^6$ m, $v = 150$ m/s.

61b. $T = \lambda/v$.

61c. $f = 1/T$.

62a. $m = \rho A h$, where m is the mass of the water wall, ρ is the density of seawater, h is the height and A is the area of the water wall. $K = mv^2/2$.

62b. $U_g = mgh/2$, $h/2$ is the height of the center of mass of the wall.

62c. $p = mv$.

63. $\Delta t = s/v + (s' + s'')/c$, where v is the speed of sound, c is the speed of light, $s = 2$ km, $s' = 1000$ km, $s'' = 5000$ km.

64a. $\lambda = v/f$.

64b. $T = 1/f$.

64c. $L = v\tau$, where τ is the duration of the pulse, and v is the speed of sound in water.

64d. $N = L/\lambda$, where N is the number of wavelengths in the pulse.

65. x-component of the total angular momentum must be conserved. Thus $L_x = -\hbar/2$.

66. $\vec{\mathbf{p}}(\bar{\nu}_\mu) = -\vec{\mathbf{p}}(\bar{\nu}_\tau)$.

67. Distance $d = ct$, where c is the speed of light.

68. From the conservation of momentum: $2p \cos\theta = mv$.

69. $P = E/t$.

70. $v = s/t$.

71. $mv^2/2 = E$.

72. a) $M_1 r_1 = M_2 r_2$ and $r_1 + r_2 = r$, where r is the distance between the two stars, M_1 and M_2 are the masses of the stars, r_1 and r_2 are the distances between the corresponding star and the center of mass.
b) $M_1 v_1^2 / r_1 = GM_1 M_2 / r^2$,
c) $L = I_1(v_1/r_1) + I_2(v_2/r_2)$,
$I_1 = (M_1 r_1^2), I_2 = (M_2 r_2^2)$,
d) $K = \frac{1}{2}(M_1 v_1^2 + M_2 v_2^2)$.

73. See hint to solution for Problem 72.

74. Find the mechanical energy of Columbus at the surface of the Earth:

$$E = -GmM_e/R_e,$$

where m is the mass of Columbus, M_e and R_e are the mass and radius of the Earth. Then find the mechanical energy of Columbus at the height of the orbit:

$$E' = -GmM_e/(h + R_e) + (1/2)mv^2 = -(1/2)GmM_e/(h + R_e),$$

where h is the height of Columbus above the ground. The energy needed can be estimated as the difference $(E' - E)$.

75. a) $F = GmM_e/r^2$, where $r = 1.5$ Gm.
b) $F = GmM_s/(r_{es} + r)^2$, where r_{es} is the Earth–Sun distance.

76. $K = mv^2/2$.

77. $\omega = v/R$, where R is the radius of a wheel.

78. $t = v/a$ where $a = 15g/2$.

79. $F = GM_s m/r^2$, where m is the mass of the sail and r is the sail–Venus distance.

80. $v = p/m$.

81. First, find the required length L of the mini-accelerator. Then, estimate the time: $t = L/c$.

82. Estimate the angular separation of the planet and the star.

83. See Hints to solution for Problem 72a.

84. See Hints to solution for Problem 72a.

85. The planet and the star are very far away compared to their mutual separation, so each subtends an essentially constant full angle of $2R/D$, where R is the radius of the object and D is the distance to the system. The fraction of the star's surface obscured during a transit is then $4\pi r_p^2/4\pi R_s^2 = (r_p/R_s)^2$.

86. To transit, the observer must fall within an angular zone above or below the orbital plane equal to the planet's orbital radius R_o divided by the radius of the star R_s. Compared to all possible directions of view this fraction is $2(R_o/R_s)2\pi/4\pi = 0.5\%$

87. $v_{av} = \Delta r/\Delta t$.

88. On the Earth the weight of the probe of mass m is

$$w_e = \frac{GM_e m}{R_e^2}.$$

On the comet of mass M_c and radius R_c the weight is

$$w_c = \frac{GM_c m}{R_c^2}.$$

Thus,

$$\frac{w_c}{w_e} = \frac{M_c}{M_e} \cdot \left(\frac{R_e}{R_c}\right)^2 .$$

From this equation one can find the ratio M_c/M_e.

7.2. Electricity and Magnetism

1. $E = k_c|q|/r^2$, where $q = 2e/3$.

2. $V = k_c q/r$.

3. $F = k_c|q_1 q_2|/r^2$, where $q_1 = 2e/3$, $q_2 = -e/3$.

4. Let the x-axis point in the direction of the electric field. The electric force on the tau lepton has only an x-component, $F_x = qE_x = -eE$. The acceleration of the tau lepton is $a_x = -eE/m$. The displacement has only the x-component

$$\Delta x = v_{xi}t + \frac{1}{2}a_x t^2 ,$$

where $v_{xi} = v_i = 10$ km/s, and the time t is taken to be the mean lifetime of the tau lepton.

5. The same as in problem 4, but $v_{xi} = -v_i$.

6. $C = \epsilon_o A/s$, $V = Q/C$, $E = V/s$, $F = qE = eE$, $a = F/m$. The direction is from the negative plate to the positive plate.

7. $F = qvB$, $q = e$, $a = F/m$.

8. The magnetic force is perpendicular to the velocity. Thus, the muon moves in a circular orbit at constant speed v. The mean distance traveled is vt, where t is the mean lifetime.

9. $U = (k_c/r)(q_1 q_2 + q_1 q_3 + q_2 q_3)$ where q_i are the electric charges of the quarks.

11. $F = k_c e^2/r^2$, $\quad a = F/m$.

13. $E = k_c q/r^2$, where $q = 2e$.

14. $V = k_c q/r$.

15. $^{208}_{82}Pb$ is the stable isotope with the maximum electric charge. Thus, $F_{\max} = k_c q^2/r^2$ where $q = 82e$.

16. $\Delta U = 2mB$, where $m = 2.1\mu_n$.

17. $\tau_{\max} = mB$.

18. $B = 2m/r^3$, where $r = 1$ nm, positive z-direction.

19a. $B = 2m/r^3$, positive z-direction.

19b. $B = m/r^3$, negative z-direction.

20. $F = k_c e^2/r^2$.

21. $k_c e^2/r$.

22. $eV = K$, where $V = 1$ MV.

23. Assume that the mass of the fireball is approximately twice the mass of a lead nucleus.

$$\rho_1 = m/V_1, \quad \rho_2 = 2m/V_2,$$

$$\rho_2/\rho_1 = 2V_1/V_2 = 2(r_1/r_2)^3 = 20, \quad r_2 = r_1/(10)^{\frac{1}{3}},$$

where m is the mass of the lead nucleus, V_1 is its volume, V_2 is the volume of the fireball. Next,

$$E_1 = k_c q/r_1^2, \quad E_2 = 2k_c q/r_2^2, \quad E_2/E_1 = 2(r_1/r_2)^2,$$

where E_1 is the magnitude of the electric field near the lead nucleus, E_2 is the magnitude of the electric field near the fireball, and q is the electric charge of the lead nucleus.

24. $B = 2m/r^3$, where m is the magnetic moment, $r = 1$ μm.

25. $B_{\max} = 2m_p/(r_o - A)^3$, $\quad B_{\min} = 2m_o/(r_o + A)^3$, $\quad \Delta B = B_{\max} - B_{\min}$, where m_o is the magnetic moment of the magnet, r_o is the equilibrium spacing between the magnet and the atom, and A is the amplitude of the cantilever vibrations.

26a. $m_\mu^{(e)} - m_\mu^{(t)} \approx m_\mu^{(t)} \times 1.3 \times 10^{-6}$ where $m_\mu^{(e)}$ and $m_\mu^{(t)}$ are the experimental and theoretical values of m_μ.

26b. $\Delta U^{(t)} = 2m_\mu^{(t)} B$, $\quad \Delta U^{(e)} = 2m_\mu^{(e)} B$,

$$\Delta U^{(e)} - \Delta U^{(t)} = 2B(m_\mu^{(e)} - m_\mu^{(t)}).$$

27. As shown in Fig. 16 below, the magnetic fields reinforce each other at the bottom and cancel each other at the top.

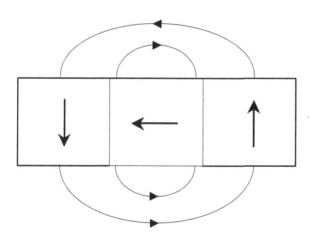

Fig. 16.   Magnetic field lines near the Halbach array.

28. Consider the current $I = 2$ pA in the gap of length 2 nm.

$$\vec{B}_p \approx \frac{\mu_o}{4\pi} \frac{I\Delta\vec{L} \times \vec{n}_p}{r_p^2},$$

where $\Delta\vec{\mathbf{L}} = -2 \times 10^{-9}\vec{\mathbf{k}}$ (in m), $\vec{\mathbf{n}}_p = \vec{\mathbf{i}}x_p/r_p + \vec{\mathbf{j}}y_p/r_p$, $r_p = (x_p^2 + y_p^2)^{\frac{1}{2}}$.

29a. $\vec{\tau} = \vec{\mathbf{m}} \times \vec{\mathbf{B}}$, where $\vec{\mathbf{m}} = \vec{\mathbf{i}}\mu_B$.

29b. $\Delta U = (\vec{\mathbf{m}}_f - \vec{\mathbf{m}}_i)\vec{\mathbf{B}}$, where $\vec{\mathbf{m}}_i = \vec{\mathbf{i}}\mu_B$, $\vec{\mathbf{m}}_f = -\vec{\mathbf{i}}\mu_B$, $\vec{\mathbf{m}}_i$ and $\vec{\mathbf{m}}_f$ are the "initial" and "final" magnetic moments, i.e. the magnetic moments before and after reversal.

30. $Q = CV$.

31. $Q = I\Delta t$, $I = jA$, where j is the electric current density and Q is the charge passing through the cross-sectional area A.

32a. $\vec{\mathbf{B}} \approx \dfrac{\mu_o}{4\pi}\dfrac{I\Delta\vec{\mathbf{L}} \times \vec{\mathbf{n}}_p}{r_p^2}$, where $I = jA$ is the current in the wire, $\Delta\vec{\mathbf{L}} = \vec{\mathbf{k}}\Delta L$, ΔL is the length of the segment and the components of $\vec{\mathbf{n}}_p$ are given by $n_{px} = x/r_p$, $n_{py} = y/r_p$, $n_{pz} = z/r_p$, $r_p = (x_p^2 + y_p^2 + z_p^2)^{\frac{1}{2}}$.

32b. $\vec{\mathbf{F}} = q\vec{\mathbf{v}} \times \vec{\mathbf{B}} = -e\vec{\mathbf{v}} \times \vec{\mathbf{B}}$, where $\vec{\mathbf{v}} = \vec{\mathbf{k}}v$.

33a. $m = IA_o = jAA_o$, where $A_o = \pi r_o^2$ is the area of the current loop (do not confuse it with the cross-sectional area of the wire A), r_o is the radius of the loop.

33b. $\vec{\mathbf{B}}_p = \dfrac{\mu_o}{4\pi}\dfrac{3(\vec{\mathbf{m}} \cdot \vec{\mathbf{n}}_p)\vec{\mathbf{n}}_p - \vec{\mathbf{m}}}{r_p^3}$, where $\vec{\mathbf{m}} = \vec{\mathbf{k}}m$ is the magnetic moment of the current loop. The components of $\vec{\mathbf{n}}_p$ are given in the solution to problem 32a.

33c. $\vec{\tau}_a = \vec{\mathbf{m}}_a \times \vec{\mathbf{B}}_p$, where $\vec{\tau}_a$ is the torque on the atom and $\vec{\mathbf{m}}_a = \vec{\mathbf{k}}m_a = 2\mu_B\vec{\mathbf{k}}$ is the atomic magnetic moment.

34a. $|\Delta U| = 2Bm$, where ΔU is the change of the magnetic energy, $B = 1\,T$, $m = IA_o = jAA_o$, A is the cross-sectional area of the wire and A_o is the area of the square loop.

34b. $\tau = 0$ as the magnetic moment of the current loop is parallel to the magnetic field.

35. The magnetic field (in T) for ion number n is equal to $B_n = 5 + 5 \times 10^{-4}(n - 1)$. For the left ion, $n = 1$, for its neighbor $n = 2$, and for the right end ion $n = 100$. The energy difference in each case is $\Delta E_n = 2\mu_B B_n$.

36a. $\vec{\mathbf{B}}_p = \dfrac{\mu_o}{4\pi} \dfrac{3(\vec{\mathbf{m}} \cdot \vec{\mathbf{n}}_p)\vec{\mathbf{n}}_p - \vec{\mathbf{m}}}{r_p^3}$. Let both the center of the magnet and the selected ion lie on the z-axis with the origin at the center of the magnet (see Fig. 13 for the problem). Then

$$\vec{\mathbf{m}} = \vec{\mathbf{k}}m, \quad \vec{\mathbf{n}} = -\vec{\mathbf{k}}, \quad \vec{\mathbf{B}}_p = \frac{\mu_o}{4\pi} \frac{2m}{r_p^3}\vec{\mathbf{k}},$$

where $\mu_o m/v = 2.2$ T, $V = 4\pi r^3/3$, r is the radius of the magnet, and r_p is the distance between the center of the magnet and the selected ion.

36b. $\tau_{\max} = \mu_B B_p$, where μ_B is the magnetic moment of the ion, and B_p is found in (a).

36c. $\Delta E_{\max} = 2\mu_B B_p$.

37. The magnetic field on the chain has only the z-component (see Fig. 14). Thus, $\vec{\mathbf{B}} = \vec{\mathbf{k}}B$ for $x > 0$, $\vec{\mathbf{B}} = -\vec{\mathbf{k}}B$ for $x < 0$, and $\vec{\mathbf{B}} = 0$ for $x = 0$. For $x = \pm a$,

$$B = B_1 - B_2, \quad B_1 = \frac{\mu_o I}{2\pi r_1}, \quad B_2 = \frac{\mu_o I}{2\pi r_2},$$

where r_1 is the distance between the atom and the center of the nearest wire, r_2 is the similar distance to the second wire, $r_1 = d/2 + \Delta/2 - a$, $r_2 = d/2 + \Delta/2 + a$.

38. $U(0) - U(a) = 3\mu_B B(a)$, $U(a) - U(2a) = 3\mu_B[B(a) - B(2a)]$,

$$U(0) - U(2a) = [U(0) - U(a)] + [U(a) - U(2a)],$$

where $U(0)$, $U(a)$, and $U(2a)$ are the magnetic energies at the positions $x = 0$, $x = a$, $x = 2a$, and $B(a)$ and $B(2a)$ are the magnetic fields (magnitude) at $x = a$ and $x = 2a$. Magnetic energy decreases with increasing x, for $x > 0$.

39. $E = P\Delta t$, where P is the power, $\Delta t = 1$ year.

40. $I_r = P/4\pi r^2$, where P is the power radiated by the Sun and r is the distance from the center of the Sun.

41. $P_r = I_r/c$. Here I_r is the intensity of the Sun's radiation $I_r = P/4\pi r^2$, P is the light power generated by the Sun and r is the distance between the center of the Sun and the asteroid.

42. $F = evB$, where v is the electron speed and B is the magnitude of the magnetic field.

43. $\tau = \mu_B B$, where B is the magnitude of the magnetic field.

44a. $\lambda = c/f$.

44b. $T = 1/f$.

44c. $L = c\tau$, where τ is the duration of the pulse.

44d. $N = L/\lambda$.

45a. $E_i \times (0.7)^n < 0.1 E_i$ where E_i is the initial energy of the pulse and n is the number of reflections. From this equation we get $n > 6.4$, consequently, $n = 7$.

45b. The distance is about $7s$, where s is the separation between the slabs.

45c. $\Delta t \approx 7s/c$.

46. $s = \lambda/2$, where s is the distance between two neighboring traps, λ is the wavelength of the light, $\lambda = c/f$, and f is the frequency of the light.

47a. $L = c\tau$.

47b. $L = c'\tau$, where $c' = 900$ m/s is the speed of the signal light in the gas.

48. $N = s/\lambda$, where N is the number of wavelengths and s is the distance between the Small Magellanic Cloud and the Sun.

49. The distance to the cluster can be estimated as its size divided by its angular size, which is 10^{10} ly. Thus, the x-rays were emitted 10 billion years ago.

50a. $\lambda = c/f$

50b. $2s$. (See Fig. 17.)

50c. $2w$. (See Fig. 17.)

51. $f = c/\lambda$.

52. $T = \lambda/c$.

53. Find the speed of a proton of energy $E = 1$ eV: $E = mv^2/2$. Then find the magnetic force: $F = evB$.

54. $d = ct$.

55. $F = (2I/c)A$, where I is the radiation intensity, and A is the solar sail area.

56. $I_e/I_u = (r_{us}/r_{es})^2$, where I_e and I_u are the intensities at the

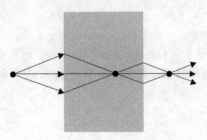

Fig. 17.   Images of the point source inside and outside the slab.

Earth and Uranus orbits and r_{us}/r_{es} are the Uranus–Sun and Earth–Sun distances.

57. a) $u_E = \epsilon_0 E^2/2$,
    b) $F = eE$.

58. a) The average intensity I_r equals half of the peak intensity. The average energy density $u_{av} = I_r/c$.
    b) $P_r = I_r/c$,
    c) Use $u_{av} = \epsilon_0 E_0^2/2$, $B_0 = E_0/c$.

59. Use $F = ecB$.

60. Power $P = IV$. Energy $E = Pt$.

Chapter 8

Conclusion

The purpose of our book is to bridge the gap between the typical introductory course of undergraduate physics and the frontiers of physics research. We believe that both undergraduate education and contemporary science would greatly benefit from such a reconciliation. We have chosen one way to approach the goal: to combine in one book selected topics from contemporary science with problems encountered in frontier research that can be solved within the scope of the "standard" undergraduate course. We believe university professors will use our problems in their courses. We also hope the book will be interesting to undergraduate and graduate students who are curious about contemporary scientific research.

Chapter 9

Appendices

9.1. Some Formulas from Vector Analysis

$$\vec{A} \cdot \vec{B} = A_x B_x + A_y B_y + A_z B_z \tag{9.1}$$

$$\vec{A} \cdot \vec{B} = \vec{B} \cdot \vec{A} \tag{9.2}$$

$$\vec{A} \cdot (\vec{B} + \vec{C}) = \vec{A} \cdot \vec{B} + \vec{A} \cdot \vec{C} \tag{9.3}$$

$$\vec{A} \cdot \vec{B} = AB \cos \phi \tag{9.4}$$

$$\vec{i} \times \vec{i} = \vec{j} \times \vec{j} = \vec{k} \times \vec{k} = 0 \tag{9.5}$$

$$\vec{i} \times \vec{j} = \vec{k}, \quad \vec{j} \times \vec{k} = \vec{i}, \quad \vec{k} \times \vec{i} = \vec{j} \tag{9.6}$$

$$\vec{A} \times \vec{B} = -\vec{B} \times \vec{A} \tag{9.7}$$

$$\vec{A} \times (\vec{B} + \vec{C}) = \vec{A} \times \vec{B} + \vec{A} \times \vec{C} \tag{9.8}$$

$$|\vec{A} \times \vec{B}| = AB \sin \phi \tag{9.9}$$

ϕ is the smaller of the angles between the vectors and $|\ldots|$ means the magnitude of the enclosed vector.

9.2. Abbreviations for Units

A	ampere
AU	astronomical unit
C	coulomb
d	day
eV	electron volt
F	farad
ft	foot, feet
g	gram
h	hour
Hz	hertz
J	joule
K	kelvin
L	liter
lb	pound
ly	light-year
m	meter
mi	mile
min	minute
mph	miles per hour
N	newton
rad	radian
rev	revolution
s	second
T	tesla
V	volt
W	watt
y	year
yd	yard
°C	degree celsius

9.3. Prefixes

$$\text{exa (E)} = 10^{18}$$
$$\text{peta (P)} = 10^{15}$$
$$\text{tera (T)} = 10^{12}$$
$$\text{giga (G)} = 10^{9}$$
$$\text{mega (M)} = 10^{6}$$
$$\text{kilo (k)} = 10^{3}$$
$$\text{centi (c)} = 10^{-2}$$
$$\text{milli (m)} = 10^{-3}$$
$$\text{micro } (\mu) = 10^{-6}$$
$$\text{nano (n)} = 10^{-9}$$
$$\text{pico (p)} = 10^{-12}$$
$$\text{femto (f)} = 10^{-15}$$
$$\text{atto (a)} = 10^{-18}$$

9.4. Some Conversion Factors

$$1 \text{ AU} = 1.496 \times 10^{11} \text{ m}$$
$$1 \text{ d} = 86400 \text{ s}$$
$$1 \text{ eV} = 1.602 \times 10^{-19} \text{ J}$$
$$1 \text{ ft} = 0.3048 \text{ m}$$
$$1 \text{ h} = 3600 \text{ s}$$
$$1 \text{ L} = 10^{-3} \text{ m}^3$$
$$1 \text{ lb} = 4.448 \text{ N}$$
$$1 \text{ ly} = 9.46 \times 10^{15} \text{ m}$$
$$1 \text{ MeV}/c^2 = 1.78 \times 10^{-30} \text{ kg}$$
$$1 \text{ mi} = 1609 \text{ m}$$
$$1 \text{ y} = 3.156 \times 10^{7} \text{ s}$$
$$1 \text{ yd} = 0.9144 \text{ m}$$

9.5. Approximate Values for Some Physical Quantities

$$\epsilon_o = 8.854 \times 10^{-12} \text{ N/V}^2$$
$$\mu_B = 9.274 \times 10^{-24} \text{ Am}^2$$
$$\mu_n = 5.051 \times 10^{-27} \text{ Am}^2$$
$$\mu_o = 4\pi \times 10^{-7} \text{ N/A}^2$$
$$\rho_{\text{air}} = 1.3 \text{ kg/m}^3$$
$$\rho_{\text{ice}} = 917 \text{ kg/m}^3$$
$$\rho_{\text{seawater}} = 1025 \text{ kg/m}^3$$
$$c = 2.998 \times 10^8 \text{ m/s}$$
$$e = 1.602 \times 10^{-19} \text{ C}$$
$$G = 6.67 \times 10^{-11} \text{ Nm}^2/\text{kg}^2$$
$$g = 9.8 \text{ m/s}^2$$
$$h = 6.626 \times 10^{-34} \text{ Js}$$
$$\hbar = 1.055 \times 10^{-34} \text{ Js}$$
$$k_B = 1.38 \times 10^{-23} \text{ J/K}$$
$$k_c = 8.99 \times 10^9 \text{ V}^2/\text{N}$$
$$m_e = 9.11 \times 10^{-31} \text{ kg}$$
$$m_p = 1.67 \times 10^{-27} \text{ kg}$$
$$N_A = 6.02 \times 10^{23}$$

9.6. Some Astronomical Data

Sun
Mass: 1.99×10^{30} kg
Radius: 6.96×10^8 m

Mercury
Mass: 3.28×10^{23} kg
Radius: 2.57×10^6 m

Semimajor Axis: 0.387 AU
Period: 88.0 d

Venus
Mass: 4.82×10^{24} kg
Radius: 6.31×10^6 m
Semimajor Axis: 0.723 AU
Period: 225 d

Earth
Mass: 5.97×10^{24} kg
Radius: 6.38×10^6 m
Semimajor Axis: 1.00 AU
Period: 1.00 y

Mars
Mass: 6.42×10^{23} kg
Radius: 3.43×10^6 m
Semimajor Axis: 1.52 AU
Period: 1.88 y

Jupiter
Mass: 1.89×10^{27} kg
Radius: 7.18×10^7 m
Semimajor Axis: 5.20 AU
Period: 11.9 y

Saturn
Mass: 5.69×10^{26} kg
Radius: 6.03×10^7 m
Semimajor Axis: 9.54 AU
Period: 29.5 y

Uranus
Mass: 8.66×10^{25} kg
Radius: 2.67×10^7 m
Semimajor Axis: 19.2 AU

Period: 84.0 y

Neptune
   Mass: 1.03×10^{26} kg
   Radius: 2.48×10^7 m
   Semimajor Axis: 30.1 AU
   Period: 165 y

Pluto
   Mass: 1.1×10^{22} kg
   Radius: 4×10^5 m
   Semimajor Axis: 39.5 AU
   Period: 249 y

Moon
   Mass: 7.35×10^{22} kg
   Radius: 1.74×10^6 m
   Average Earth–Moon distance: 3.84×10^8 m
   Period: 27.3 d

9.7. Notation Used

$\vec{\alpha}$	angular acceleration
β	sound level
β^-	β^--particle (electron)
β^+	β^+-particle (positron)
γ	photon
$\vec{\Delta\theta}$	angular displacement
ϵ_o	permittivity of free space
ϵ	permittivity of an insulating substance
$\vec{\theta}$	angular position
λ	wavelength
μ^-	muon
μ^+	antimuon
μ_o	permeability of free space
μ_B	Bohr magneton

μ_k	coefficient of kinetic friction
μ_n	nuclear magneton
μ_s	coefficient of static friction
ν_μ	muon neutrino
$\bar{\nu}_\mu$	muon antineutrino
ν_τ	tau neutrino
$\bar{\nu}_\tau$	tau antineutrino
ν_e	electron neutrino
$\bar{\nu}_e$	electron antineutrino
π^+	pion
π^-	antipion
$\vec{\tau}$	torque
ρ	density
ρ	resistivity
σ	electric surface charge density
τ^-	tau lepton
τ^+	tau antilepton
ϕ	angle between two vectors
ϕ	phase constant
ϕ_e	electric flux
ϕ_m	magnetic flux
$\vec{\omega}$	angular velocity
ω	angular frequency
A	amplitude
A	area
$\vec{a}$	acceleration
a_c	magnitude of centripetal component of acceleration
a_t	magnitude of tangential component of acceleration
$\vec{B}$	magnetic field
b	bottom quark
$\bar{b}$	bottom antiquark
C	capacitance
c	speed of light
c	charm quark
$\bar{c}$	charm antiquark
d	down quark

$\bar{d}$	down antiquark
E	mechanical energy
E_m	magnetic energy
$\mathcal{E}$	electromotive force (EMF)
e	fundamental charge
e^-	electron
e^+	positron(antielectron)
$\vec{\mathbf{E}}$	electric field
$\vec{\mathbf{F}}$	force
$\vec{\mathbf{F}}_g$	gravitational force
$\vec{\mathbf{F}}_n$	normal force
$\vec{\mathbf{F}}_k$	kinetic friction force
$\vec{\mathbf{F}}_s$	static friction force
$\vec{\mathbf{F}}_T$	tension force
f	frequency
G	universal gravitational constant
g	gravitational acceleration at the Earth's surface
h	height above the surface of the Earth
h	Planck's constant
$\hbar$	$h/2\pi$
$\vec{\mathbf{I}}$	impulse
I	electric current
I	intensity
I	moment of inertia
I_o	threshold of hearing
$\vec{\mathbf{i}}$	unit vector in the $+ x$-direction
$\vec{\mathbf{j}}$	unit vector in the $+ y$-direction
j	electric current density
$\vec{\mathbf{k}}$	unit vector in the $+ z$-direction
K	kinetic energy
k	wave number
k_B	Boltzman's constant
k_c	Coulomb's constant
k_s	spring or force constant
$\vec{\mathbf{L}}$	angular momentum

L	length
L_m	inductance
$\vec{\mathbf{m}}$	magnetic moment
M_e	mass of the Earth
M_o	mass of the Sun
m	mass
m_e	electron mass
m_p	proton mass
N_A	Avogadro's number
n	refractive index
n	neutron
$\bar{n}$	antineutron
$\vec{\mathbf{p}}$	momentum
P	power
P_r	radiation pressure
p	proton
$\bar{p}$	antiproton
Q	electric charge on a capacitor
q	electric charge
$\vec{\mathbf{r}}$	position vector
R	resistance
R_e	radius of the Earth
R_o	radius of the Sun
$\vec{\mathbf{S}}$	Poynting vector
S	surface
s	distance traveled by a particle
s	separation of capacitor plates
s	strange quark
$\bar{s}$	strange antiquark
T	period
t	top quark
$\bar{t}$	top antiquark
U	potential energy
U_g	gravitational potential energy
U_s	spring potential energy
u	up quark

$\bar{u}$	up antiquark
u_E	electric energy density
u_B	magnetic energy density
V	volume
V	voltage
V_P	electric potential at the point P
$\vec{v}$	velocity vector
$\vec{v}_{av}$	average velocity
W	work

Index

Printed in the United States
By Bookmasters